THE ART OF TEACHING PRIMARY SCIENCE

'This book includes the wisdom of very many of Australia's leading experts in the field of science education in the primary years. I commend it as an excellent reference for those teachers of the primary years seeking the best ways to engage their students in good science and scientific investigation, and keen to link these with other learning areas such as literacy, technology and numeracy. I particularly commend it to those teachers who are embracing the Australian Academy of Science *Primary Connections* professional learning program, where it will provide most valuable and detailed background and support.'

Peter Turnbull, President, Australian Science Teachers Association

THE ART OF TEACHING PRIMARY SCIENCE

EDITED BY VAILLE DAWSON
AND GRADY VENVILLE

ALLEN&UNWIN

First published in 2007

Allen & Unwin
83 Alexander Street
Crows Nest NSW 2065
Australia
Phone: (61 2) 8425 0100
Fax: (61 2) 9906 2218
Email: info@allenandunwin.com
Web: www.allenandunwin.com

National Library of Australia
Cataloguing-in-Publication entry:

  The art of teaching primary science.

  Bibliography.
  Includes index.
  ISBN 978 1 74175 289 2 (pbk.).

  1. Science - Study and teaching (Primary). I. Dawson,
  Vaille Maree. II. Venville, Grady Jane.

  507.1

Set in 11/13 pt Legacy Serif Book by Midland Typesetters, Australia
Printed and bound in Australia by Griffin Press

10 9 8 7 6 5 4 3 2 1

CONTENTS

FIGURES AND TABLES

FIGURES

TABLES

CONTRIBUTORS

Peter Aubusson is Associate Professor in the Faculty of Education at the University of Technology, Sydney, where he leads the Designs for Learning Research Cluster. He was a teacher for ten years before becoming a university teacher and researcher. He has been researching science education for more than 15 years.

Coral Campbell started her professional life as a scientist before becoming a primary teacher. Since joining Deakin University Coral has been involved in primary and secondary science, technology and mathematics education. She works closely with local schools, promoting experiential learning for her education students by placing them in schools.

Marj Colvill, a Tasmanian early childhood teacher, has worked for the past 30 years to support primary science education and has made a significant contribution to primary science education through her roles in the Science Teachers Association of Tasmania and the Australian Science Teachers Association. In 2006 she was awarded the Prime Minister's Award for Excellence in Primary Science Teaching.

Ian Ginns is a tertiary science educator interested in the effective teaching of science so that students benefit from active engagement in the subject. Recently, he has investigated how integrating the teaching of science and technology can enhance these benefits for students, while maintaining the unique identity of each learning area.

Denis Goodrum is Professor and Head of the School of Education and Community Studies at the University of Canberra. He has a national and international reputation in science education and has provided leadership for a number of national projects of significance.

Janette Griffin previously taught students from K to 12 in formal and informal settings and is now Senior Lecturer in Science Education and Learning Beyond the Classroom at the University of Technology, Sydney. Her research and publications investigate ideal conditions for integrated school/museum learning and the complementary roles of teachers and museum educators.

Mark Hackling is Professor of Science and Technology Education at Edith Cowan University. Mark has led a number of national science education initiatives and is currently Co-director of the *Primary Connections* Project. Mark has research interests in teacher professional learning, and improving investigation work and assessment in school science.

Allan Harrison is Associate Professor at Central Queensland University and teaches science methods to primary and secondary pre-service teachers. Allan's research interests include the nature of science, analogies, metaphors and models, and conceptual change. He is on the board of the National Association for Research in Science Teaching (NARST: a US-based organisation) and has published in the leading science education journals.

Ruth Hickey is a Senior Lecturer at James Cook University, Cairns. She taught in schools throughout Western Australia, and has contributed to projects in the English and Science curricula. Her research focus is conceptual development of science concepts and, in particular, how this impacts on pre-service and classroom teachers' decision-making when supporting student progress.

Christine Howitt is a Lecturer at Curtin University of Technology in Western Australia. Her role involves postgraduate teaching and supervision within science education. Her research interests are in primary science teacher education with a specific focus on developing positive attitudes towards and increased confidence in science in pre-service primary teachers.

Mary Morris is a Lecturer in Science Education at Edith Cowan University. She has a teaching background, with 20 years' teaching in early childhood and primary education both in Australia and in the United Kingdom. Mary is an active member of the Science Teachers Association of Western Australia and has a keen interest in promoting science to the wider community. She is currently co-editor of SCIOS, the journal of the Science Teachers Association of WA.

Steve Norton began his educational vocation in 1983 as a secondary school teacher of mathematics and science. Since completing his PhD in 2000 he has lectured in primary and secondary mathematics and science education. His research focus has included the use of information communication technologies (ICTs) and technology in learning, integrated learning and algebra learning.

Jennifer Pearson's teaching career has spanned primary, secondary and tertiary education positions. She has also worked with landcare groups to support schools as they engage in Environmental Education projects. Jennifer is involved in the Science Teachers Association as the journal co-editor and is the Convenor of the Australian Association for Environmental Education (AAEE) Chapter in WA.

Vaughan Prain is Professor and Head of the School of Education at La Trobe University in Victoria. Combining his expertise in science and literacy, his research interests are in literacy education and new technologies for learning. He is the Literacy Consultant for the *Primary Connections* Project and co-author of *Writing and Learning in the Science Classroom* (Kluwer, 2004).

Christine Preston is a Lecturer in primary curriculum studies including Science Education at the University of Sydney. She also teaches kindergarten science at Abbotsleigh Junior School. Christine lectured in early childhood, primary and secondary science teacher education at Macquarie University and taught secondary and K–12 science before specialising in primary science.

Stephen Ritchie is an Associate Professor in Science Education at Queensland University of Technology. He is co-author of *Re/Constructing Elementary Science* (Peter Lang, 2001) with Michael Roth and Ken Tobin, and co-editor of *Metaphor and Analogy in Science Education* (Springer, 2006) with Peter Aubusson and Allan Harrison.

Russell Tytler has been involved over many years with primary science teacher education, science curriculum development, and professional development projects including 'School Innovation in Science'. His research interests include teacher learning, student learning, reasoning and investigating in science, learning and literacy, and public understandings of science.

Wilhelmina Van Rooy is a Senior Lecturer in the School of Education at Macquarie University. Her research interests are in biology education, teacher education and children's understanding of health. She is the Chief Examiner for the NSW Higher School Certificate for 2006/07.

ABOUT THE EDITORS

Vaille Dawson is a Senior Lecturer in Secondary Science Education at Edith Cowan University. Prior to commencing an academic career Vaille worked as a medical researcher and taught secondary school science for ten years. Her research interests relate to adolescents' understandings, attitudes and decision-making about biotechnology issues and the use of information technology by early-career science teachers. She is co-editor of the textbook for pre-service secondary science teachers, *The Art of Teaching Science* (Allen & Unwin, 2004).

Grady Venville is a Professor of Science Education at the University of Western Australia. She has taught Science, English as a Second Language and Science Education in primary, secondary and tertiary institutions in Australia, England and Japan. Grady has published widely in international and national journals on curriculum integration, conceptual change, students' learning in genetics and cognitive acceleration. Grady is co-author of *Let's Think* (NFER Nelson, 2001), a curriculum package on cognitive acceleration for Year 1 teachers, and co-editor of *The Art of Teaching Science* (Allen & Unwin, 2004).

PREFACE

Young children have a most unaffected and wondrous curiosity about the natural world that makes teaching science in primary school a delightful and rewarding experience. The inquisitive questions that seem to come from nowhere, the look of delight as a child discovers something: these are the encounters that, for many teachers, transform the noble profession of teaching into a passion. This does not mean that teaching science to primary-aged children is an easy thing to do, far from it. Teaching primary science is an art that requires, among other things, knowledge of the ways that children learn about science, knowledge about nature of science, skills and expertise with a broad spectrum of effective teaching approaches, careful planning, an array of teaching resources, ingenuity and a large dose of patience. This book is about the art of teaching science to children in primary and early childhood educational settings. We have drawn together some of the most highly-regarded and well-known science educators in Australia and asked them to interpret their own understandings and experience of the art of teaching primary science.

The chapters in *The Art of Teaching Primary Science* have been carefully crafted to be accessible for primary and early childhood student teachers. Each chapter addresses a major topic of importance, but we acknowledge that this is an artificial structure. Teaching primary science is an integrated and complex professional activity. We also understand that no book can provide all that is needed to master the art of teaching primary science. To do this, pre-service teachers need a number of things including guidance and mentoring from an experienced and knowledgeable teacher educator and a well-designed primary education course with an integrated teaching practice experience. We have designed *The Art of Teaching Primary Science* to support and enrich this learning process. It can be used as an advance organiser, to prepare student teachers for each of the topics that are addressed in a pre-service teaching course or, alternatively, it can be used for the purpose of reflection and revision. Experienced teachers can use the book to improve their practice and to enable more confidence and creativity when teaching primary science.

The status and quality of primary science teaching in Australia has attracted a great deal of attention from researchers and educators in recent years (see, for example, Goodrum, Hackling & Rennie, 2001). Such research has culminated in the publication of a new and innovative primary science teaching resource called *Primary Connections* (Australian Academy of Science, 2005). *Primary Connections* is currently being implemented in many schools in Australia. The Australian Science Teachers Association and each of the member state and territory science teachers associations have been very supportive of the implementation of *Primary Connections* and involved in the delivery of the associated professional learning. Several of the chapter authors of *The Art of Teaching Primary Science* were consultants in the development of *Primary Connections*. As a consequence, *The Art of Teaching Primary Science* is consistent with the theory and practices underpinning *Primary Connections*.

The Art of Teaching Primary Science is a sequel to *The Art of Teaching Science* (2004), our secondary-focused book. The two books are similar in structure and in some of the chapter topics because we feel we have found a structure that works well for the purpose of science teacher education. The vast majority of material in this book is, however, completely new because it is written for the primary teacher and uses examples specific to primary and early childhood education contexts. We have tried to retain the features that are common to good science teaching in both primary and secondary education settings while emphasising the features that are unique to the teaching of primary science.

The book is divided into three parts. The first, 'Understanding the Art of Teaching Primary Science', provides a theoretical base for the book. Thus, readers will become aware of nature of science, students' conceptions of science and views of learning, conceptual development and the Australian state and territory science curriculum documents. The second part, 'Implementing the Art of Teaching Primary Science', provides an overview of the fundamental aspects of teaching science effectively to primary-aged children—planning, teaching strategies, investigations, resources and assessment. The third part, 'Extending the Art of Teaching Primary Science', takes readers beyond the basics and enables them to engage with and explore some of the issues facing primary science teachers. It includes a chapter on the literacies of science that articulates the ideas underpinning the curriculum package *Primary Connections*. Other chapters explore the integration of science and technology, learning science and technology outside the classroom and the unique aspects of learning science in early childhood settings.

Finally, we wish to encourage readers in their endeavours with primary science teaching. Engaging children in science at the earliest stages of their education is critical for both general scientific literacy as well as sustained interest in science-based subjects and science-based careers. We hope that this book will enable you to develop much theoretical wisdom, practical know-how and professional enthusiasm.

Vaille Dawson
Edith Cowan University, Western Australia

Grady Venville
University of Western Australia, Western Australia

REFERENCES

Australian Academy of Science. (2005). *Primary connections.* Canberra: Australian Academy of Science.

Goodrum, D., Hackling, M., & Rennie, L. (2001). *The status and quality of teaching and learning of science in Australian schools: A research report.* Canberra: Department of Education, Training and Youth Affairs.

PART I
UNDERSTANDING THE ART OF TEACHING PRIMARY SCIENCE

CHAPTER 1

THE WONDER OF SCIENCE

Allan Harrison
Central Queensland University, Queensland

...Eye

...Feeler

OUTCOMES

By the end of this chapter you will:
- understand that science knowledge is created by people and changes as people discover new objects and processes;
- understand that scientists think about problems and develop theories that make predictions that can be tested; and
- understand that scientific theories are tested, changed and retested until every conceivable objection has been dealt with.

INTRODUCTION

One warm Queensland morning, as a teacher and I crossed a bridge over a shallow stream we saw three or four schools of tadpoles swimming in the water. The tadpoles seemed to prefer certain parts of the shallow water, and they did not respond to slow-moving shadows but darted away when a shadow quickly crossed their path. The water was flowing and clear, and little food was visible. The first questions that came to mind were: Are they cane toad or native frog tadpoles? How can we find out? As we talked about studying tadpoles and frogs in primary science, further questions arose: Why do they like to swim in some places but not others? What eats them and what do they eat? How fast do they grow? Do they sleep? What special behaviours do they show (i.e. response to light/shade, warm/cold, predators, etc.)?

As we talked, we realised that this was an excellent topic for primary science. The questions were interesting, eminently doable and open-ended. The topic is relevant because the spread of cane toads in northern Australia has major implications for native wildlife, agriculture and tourism. Studying tadpoles also is interesting because of the sharp decline in native frog numbers in Australia. Is this due to global warming, habitat destruction, pollution or new diseases? What, if anything, can we do to reverse this trend? To answer these questions requires scientific knowledge. Schools now have improved access to considerable amounts of information and students can design their own investigations, collect data and expand their knowledge of frogs, toads and habitats. A tadpole study topic has sufficient dimensions—identification, habitat study, behaviours, life cycles and population decline—to keep a class busy and such a study has the added benefit of modelling the way scientists work in groups and share and debate their findings. Individual schools could even join a state or national group that studies frog numbers and their distribution.

SCIENCE AND QUESTIONING

The tadpole study topic shows the importance of thinking and working scientifically. Questions are an excellent place to start in science. Questions draw our attention to what is already known about the subject (laws, theories, relationships and processes) and this helps in two ways: first, we don't 'reinvent the wheel' and, second, we can see what previous studies have found out and what we still need to establish. Science has more questions than we could ever describe, and all sorts of scientists are trying to solve them. As we can see in the next example, problems are the lifeblood of science.

Two Perth scientists, Barry Marshall and Robin Warren, received the 2005 Nobel Prize for Physiology and Medicine ('Australians win Nobel', 2005) for showing that stomach ulcers were not the result of a high-stress lifestyle but were, in fact, caused by bacteria. Millions of sufferers can now be cured by a course of antibiotics instead of having to take antacids all their life and eating bland foods. (See Snapshot 1.1.)

Plate tectonics and continental drift are theories that once were ridiculed but now are accepted as if they were never questioned. Scientific theories can develop in piecemeal fashion and some progress slowly, others very quickly. Alfred Wegener is often credited with the theory of continental drift; but at least ten scientists were involved in the development of plate tectonics and to whom does the credit belong? Examine the evidence in Snapshot 1.2 and decide who invented plate tectonics. (See http://kids.earth.nasa.gov/archive/pangaea/evidence.html and http://whyfiles.org/094quake/index.php?g=6.txt for useful additional information.) As you read the story, notice how scientists questioned and refined these theories and see how important the growth in technology was in answering key questions. Sometimes great ideas founder because they are too far ahead of their time and the technology necessary to provide the essential corroborating evidence.

SCHOOL SCIENCE

Let's now turn our attention to school science. How can the science we read about in the ulcers and plate tectonics stories help us improve primary science? First, stories are an excellent way to popularise science because they capture the wonder and progress of science for both teachers and students (Schwab, 1966; Millar & Osborne, 1998). The *Magic Schoolbus* series of books and DVDs (Scholastic Inc.) presents accurate science in ways that entertain and educate children. Second, stories show us that scientific knowledge is mostly hard

Snapshot 1.1: A cure for ulcers and a Nobel prize

In 1979, when Barry Marshall and Robin Warren were doctors at Royal Perth Hospital, they noticed that many tissue samples taken from the stomachs of ulcer patients were infected with bacteria. One previously unknown bacterium stood out as being present in many samples. It became known as *Helicobacter pylori* (*H. pylori*). What Marshall and Warren suspected—that *H. pylori* was causing the ulcers—would hurt drug companies. In the United States alone, ulcer sufferers consumed over US$4 billion worth of pills and antacids. Marshall and Warren's hypothesis was that *H. pylori* caused stomach ulcers and, if they were right, it could be killed with a course of antibiotics. Ulcers could be cured. This science in action story is the opening piece in Tobin and Dusheck's college biology textbook and they use the story to show how a problem and a discovery changed the theory of stomach disease (Tobin & Dusheck, 1998, pp. 1–4).

Solving problems is not easy: Marshall and Warren had to overcome established theories that said bacteria could never live in the stomach because it is too acidic. Add to this problem their relative inexperience and lack of international reputation. Experts wouldn't believe that the bacteria in the samples came from the stomach—perhaps the samples had been contaminated after they were removed from patients' stomachs? For this to happen, *H. pylori* would have to have been present in the lab but it was previously unknown.

In 1983, Marshall presented the 'bacteria hypothesis' at a conference in Brussels. He was brimming with confidence but lacking convincing evidence; the idea was dismissed. Back in Perth, Marshall and Warren tried to infect rats with *H. pylori* and give them ulcers, but none got sick. Another setback. Was the hypothesis valid? Finally, in 1984, Marshall did what a researcher should never have to do—he brewed up a culture of *H. pylori* and drank it. It made him feel sick, gave him very bad breath and two weeks later his stomach was inflamed. A biopsy showed that it was swarming with *H. pylori*. He infected himself and then cured himself with antibiotics. It was a dangerous experiment but he's now famous because it worked. Good science, imagination, tenacity, good luck or just hard work? Probably a mix of all these things.

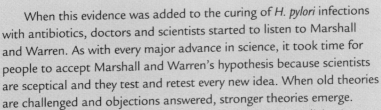

When this evidence was added to the curing of *H. pylori* infections with antibiotics, doctors and scientists started to listen to Marshall and Warren. As with every major advance in science, it took time for people to accept Marshall and Warren's hypothesis because scientists are sceptical and they test and retest every new idea. When old theories are challenged and objections answered, stronger theories emerge.

For more information, see http://nobelprize.org/medicine/laureates/2005/index.html and http://www.hpylori.com.au.

work, that scientists are a sceptical lot and that scientific theories can advance in a manner such as 'two steps forward, one step back' (sometimes one forward and two back!). Third, they help us understand that the scientists who create scientific theories are people and people who talk and argue with each other. Some theories get scientists so worked up that they stop talking to each other altogether (Latour & Woolgar, 1986). Finally, when scientists do agree on a theory such as plate tectonics, they then talk and act as if they had never thought any different. Once a problem is solved, scientists forget the trials and tribulations involved in developing the theory and mature theories are accepted as standard knowledge. We have to remember, however, that most people think and act this way and scientists are just ordinary people with extraordinary jobs.

Rather than teach a lot of science knowledge, the early years of school science should concentrate on problems that are interesting and able to be done in a primary school setting. Primary school science should excite children, encourage their curiosity and make them want to explore the world. Stories can fill knowledge gaps by providing 'need-to-know' information that is not immediately available in school but is essential for the progress of an investigation (e.g. named pictures of common frogs and tadpoles). Table 1.1 suggests a plan for school science. You may be able to see how investigating tadpoles and frogs could be adapted to the learning context, content emphasis and benefits for Years 1–3, 4–6 and 7–9, provided that the study questions are matched to the interest, capacity and needs of students in each level and local resources.

If we were to follow this model, do tadpole topics fit state and territory science syllabuses? In the Queensland syllabus

Snapshot 1.2: Who invented plate tectonics?

Back in the 1600s, English scientist and philosopher Francis Bacon noticed that the coastlines of South America and Africa fit together like a jigsaw puzzle. Two hundred years later, German scholar and explorer Alexander von Humboldt made the same observation. The first person to propose that the continents had once been joined and had moved apart was a French geographer, Antonio Snider-Pellegrini, in 1859. In 1885, Austrian geologist Eduard Suess argued that similar fossils on adjacent continents meant they once were connected. He believed that the continents moved as the molten Earth cooled and mountains formed like the wrinkles that develop on a shrivelling apple. Suess felt that southern Africa, Madagascar and India were once one continent he called Gondwana. Using his 'drying apple' model, Suess proposed that compression caused the landmass between present-day continents to move vertically downwards to form the ocean floor. In 1910, American geologist Frank Taylor stated his 'continental drift' hypothesis—he argued that moving continents pushed up mountain ranges. Most geologists, however, rejected continental drift.

In Germany, Alfred Wegener published *The Origin of Continents and Oceans* in 1915. He claimed that geological, biological and meteorological evidence combine to indicate that today's continents are fragments of an original super-continent called Pangaea. Knowing that continental and oceanic crust are radically different in density and thickness, Wegener rejected Suess's view that continental crust could be converted to oceanic crust by vertical collapse. Wegener was convinced that continental crusts had moved apart.

Geologists and geophysicists disbelieved Wegener's hypothesis because contemporary technology could not supply the necessary evidence. One problem was that ordinary volcanic activity could not account for the conduction to the surface of all the heat generated by radioactive decay in the mantle. In the 1930s, English geologist Arthur Holmes, who investigated many possible mechanisms, suggested that convection currents in a moving mantle could account for this heat loss.

Australian geologist Warren Carey then showed that the faulting and folding in New Guinea could be explained if the north-east Pacific region was moving westward relative to the Australian/PNG continental block. He proposed that ocean basins were formed when mantle material rose up as continental blocks rifted apart and left 'hot spot traces' such as the Hawaiian Islands in deep oceans.

More evidence came as technology improved. In the 1950s, surveys of magnetised rocks showed the directions they had pointed in the past. After reviewing his seafloor maps, American geologist Harry Hess discovered that lava was building mid-ocean ridges and that the continents were riding on a plastic mantle. In 1961, American geophysicist and oceanographer Robert Dietz coined the term 'seafloor spreading' to describe Hess's observations. In 1963, British geophysicists Fred Vine and Drummond Matthews published their results of a magnetic survey over part of the Carlsberg ridge in the Indian Ocean. Alternating reversely-magnetised 'stripes' showed that mantle convection currents had probably reversed the Earth's magnetic field many times in the past.

By 1965, Vine, together with Canadian Tuzo Wilson, had showed similar polarity reversals off the US coast and in 1967, Dan McKenzie and Bob Parker were the first to propose that the Earth's crust consists of a series of 'plates' which can slide, stick, break at their boundaries or plunge under another plate. Plate tectonic theory had arrived.

Finally, in 1977 hydrothermal vents or 'black smokers' (which had been predicted long before they were discovered) were found over 2500 m below the surface at the Galapagos Rift by the deep-sea submersible *Alvin*.

We now know that most volcanic activity, and earthquake epicentres, occur in the very narrow zones where tectonic plates interact. Plate tectonics is the best theory yet for explaining mid-oceanic ridges, mountain growth, volcanism and earthquakes. But will it change? Probably.

Note: I acknowledge the help of Di and Wes Nichols in providing valuable information about plate tectonics.

Table 1.1: A plan for education in science that recognises the interests, capacity and needs of school students (adapted from Fensham, 2004)

A structure for the school science curriculum		
Year	Context for learning science	Emphasis and benefit
1–3	Relevant exciting themes	Wonder about the natural world, creative thinking
4–6	Relevant everyday themes in science and technology	Concern for the natural world, conducting investigations that produce practical outcomes
7–9	Science themes with relevant applications	Motivated learning and persistent engagement that yield logical outcomes
10	Science for citizenship	Science literacies, awareness and decision-making
11–12	Optional science: disciplinary and environmental	Preparation for work and scientific wellbeing, contextually appropriate knowledge

(http://www.qsa.qld.edu.au/yrs1to10/kla/science/syllabus.html), studying tadpoles and frogs and their environment satisfies part of the Life and Living and the Science and Society strands. Fair testing is a key feature of Science and Society because hypotheses, theories and fair testing are the essence of science (see Chapter 7).

Studying tadpoles or fish, or growing plants fascinates children. Students make observations, locate expert information on the web, construct hypotheses and design ethical experiments (*in situ* and in the classroom) and sometimes their experiments work (should all experiments work? Look back at the ulcers story). Ethical issues enrich the scientific process when students have to grapple with the tension between stewardship and the avoidance of harm, and their desire to know. William Harvey's vivisection experiments on live animals led to his discovery of blood circulation (Friedman & Friedland, 1998), but would they be allowed today?

WHAT IS SCIENCE AND HOW DOES IT WORK?

The aim of science is to understand such things as quarks, atoms, microbes, plants and animals, Earth processes and how the Universe works. The full list is endless because the number of topics in science grows each year and science knowledge changes as new discoveries are made. For example, the Human Genome Project provided scientists with the knowledge, skills and technology to map the DNA of any microbe, plant or animal. A new science called genomics and phenomics was born. Genomics maps the DNA in an organism's cells (that is, it tells us exactly what the code says) and phenomics studies the ways in which an organism's genetic code appears as structures, diseases and behaviours. The Human Genome Project turned out to be more than interesting science; it turned very competitive and, occasionally, nasty. An international public project was funded by the American and British governments, among others, but Craig Venter's 'Applied Genomics'—a private project—also wanted a piece of the action. The race was on for the prize of patenting human disease genes and the resultant ownership of treatments for inherited diseases such as cystic fibrosis, diabetes and haemophilia. On the agricultural side, companies like Monsanto own genetically modified plants and animals. Trillions of dollars are at stake and Aldous Huxley's 'brave new world' may now be with us.

We may not like some of the outcomes of modern science but they are here to stay and many people now want to make informed decisions about how gene technologies, nuclear energy, therapeutic cloning, climate change and waste management affect their lives. Scientific literacy can no longer be a luxury. People want to know what science is doing and they want to have a say in government and social decisions on scientific issues. Intelligent debate requires not only scientific knowledge but the ability to collect and interpret the information that we need to make sound decisions. Science education can fill this need and science teachers need to be clear on how best to teach the concepts and facts that will help people make informed decisions. This book is a good place to start.

In summary, we do not expect to know everything in science and we do expect our current knowledge to change as technology advances. Encouraging children to imagine, ask questions, design fair tests and interpret the data that they collect about everyday issues is a goal that is both reasonable and achievable. To help you think about fundamental scientific ideas, we now examine some of the principles and processes that make science work.

Scientific literacy

People hold a variety of ideas on how science works, what qualifies as scientific knowledge and whether this knowledge is trustworthy and useful. For example, many people distrust nuclear science and the drug industry. Some are disappointed that, far from solving our energy problems, the nuclear industry has littered the environment with radioactive waste and has enabled certain countries to misuse nuclear science and build weapons of mass destruction. Others are disappointed when advertised 'wonder' drugs turn out to have lethal side effects and their distrust of science is increased when they find out that drug companies sometimes selectively report population trials and fail to publicise drug risks. It's easy to blame 'science' for these problems; however, we can also argue that people should be sufficiently educated that when they encounter conflicting information and claims, they can make informed choices. But people can only make informed judgments when they have all the relevant information, understand the concepts involved and know how to compare and contrast arguments. Thus, understanding how scientists collect and evaluate evidence is a valuable life skill.

In their *Beyond 2000* report, Millar and Osborne (1998) endorsed this idea. They argued that the primary responsibility of school science is to be 'a course to enhance general "scientific literacy"', and that this is 'an *end-in-itself*, [because science education] must provide both a good basis for lifelong learning and a preparation for life in a modern democracy' (1998, p. 2009). This is an excellent aim for science education; however, we need to ask whether the current curriculum provides both children and adults with the necessary knowledge and skills to make informed decisions in a pluralistic world. As this issue will be discussed in more detail in other chapters, I will restrict my comments here to how science should work for people in a modern world.

A survey of science principles

Science is a way of thinking about the world and the primary aim of science education is to help people think and work scientifically. A useful place to start is the aims of science and what you think it can and can't do. Step back and ask yourself: What do I think science is and what are we allowed to do in science? A shortened version of a recent science survey called 'How I Think Science Works' (HITS-W; Harrison, 2005), is presented in Table 1.2. Before you read on, read each question in Table 1.2 and tick the column that best represents your view. Please complete the survey—use a piece of scrap paper if you don't want to mark the book. Appropriate answers to some of the ideas in the survey are discussed in the paragraphs below.

Commentary on HITS-W

Theories and laws

How did you respond to the questions on theories and laws? Some people say that laws are stronger than theories but some of our best scientific understandings are called theories. For instance, the theory of relativity works everywhere (including situations where Newton's Laws of Motion break down, such as the motion of very fast space-craft and the orbit of Mercury) but it is still called a theory. The theory of relativity is more universal than Newton's Laws of Motion; so, to say that something is 'just a theory', thereby implying that a theory is somehow limited or unreliable, misses the point.

Take another example: most of us have heard of atomic theory and accept that all matter is made up of invisible particles that jiggle around (in liquids and solids) and that zip around and bounce off each other (in gases). But we've never seen individual atoms—and never expect to—despite the fact that atomic theory explains most of physics, chemistry and biology. Physicist Richard Feynman once said of this theory:

> [if] all of science knowledge were to be destroyed, and only one sentence passed on to the next generation of creatures, what statement would contain the most information . . . I believe it is the *atomic hypothesis . . . that all things are made of atoms—little particles that move around in perpetual motion, attracting each other when they are a little distance apart, but repelling upon being squeezed into one another.* (1995, p. 4)

Table 1.2: How I Think Science Works (HITS-W) (from Harrison, 2005, p. 5; reproduced by kind permission of Central Queensland University)

How I Think Science Works (HITS-W) Student version	Agree	Unsure	Disagree
1 Black holes do exist in space.			
2 Scientific laws do not have to work in every situation.			
3 People can communicate with each other using mind waves.			
4 A hypothesis is an educated guess.			
5 Scientists develop theories by doing experiments.			
6 Scientists know how my DNA determines my eye colour.			
7 Nature will reveal its secrets if we ask the right questions.			
8 Astrology is scientific.			
9 A scientific law is better than a scientific theory.			
10 One good experiment could settle a scientific argument.			
11 Scientists know that atoms exist because they have seen them.			
12 Scientists are critical of knowledge in their area of expertise.			
13 Some magicians can bend a spoon without touching it.			
14 We can use theories to make predictions that can be tested.			
15 After carrying out many successful experiments, scientists prove theories.			
16 There is life on planets outside our solar system.			
17 Once scientific laws are proved, they never change.			
18 People can do super-human things under hypnosis.			

	Agree	Unsure	Disagree
19 Some scientific theories cannot be tested.			
20 People can carry out experiments in their minds by thinking about things.			
21 Scientists are confident about forces, atoms and gravity that they cannot see.			
22 When scientists have enough evidence, theories become laws.			
23 If two repeated experiments have different results scientists are confused.			
24 A cautious hypothesis is better than a bold hypothesis.			
25 When scientific theories are tested and fail, the theory is disproved.			
26 Scientists make scientific knowledge.			
27 There is only one correct interpretation for each experiment.			

Theories don't have to be perfectly understood to be useful: they grow and change and usually improve with time as new information and evidence are discovered and they are tested in more accurate and different ways. The atomic theory (or hypothesis) is an excellent example because it explains almost everything and took 250 years to develop into the many forms it takes today. Notice that I said 'many forms', because the atomic theory we teach in primary school is refined and changed as students move through school. The many descriptions of the nature of atoms and explanations of how they interact are called 'scientific models'. Each atomic model simplifies some features and exaggerates others; this is what makes different models valid and useful and explains why we can change models to suit the problem we are dealing with.

Theories bring together a vast range of ideas, facts, experiments and evidence in ways that make sense of important processes in the natural world. Nowadays, no theory is changed into a law because there really is no difference between what used to be called a law and what is now called a theory. Certain concepts were once called laws because it was thought that they were absolutely true but, as I have explained, Einstein's general theory of relativity is more complete

than Newton's Laws of Motion. Scientists like theories because they are adaptable—they can be enhanced, tested, changed and applied to more needs and situations than we can think of—that's right, they can adapt to new situations and solve problems we haven't yet thought of. (See Snapshot 1.3 on p. 19.)

Hypotheses

Hypotheses have been called 'educated guesses' but it's better to call them 'logical statements that make testable predictions'. If you have reliable information about a natural event or object, you can predict what will happen if you change one of the factors in your hypothesis (or statement) about the event or object. Take a hypothesis about the growth of seedlings. Children know that seedlings thrive when they have plenty of light and water. But how much is enough? When we study how much water and light radish plants need to grow, for example, we cannot vary both the amount of water and the amount of light in the same experiment. If we reduce both light and water and the plant fails to thrive, we won't know if the cause was lack of light or lack of water. Fair testing tells us that we can change only one variable at a time. A hypothesis also raises the question of how to measure the effect of the variable we are going to change or manipulate. Let's settle on light and limit the light available to some of our radish plants. A useful hypothesis will say something along the lines of, 'Radish plants need bright sunlight to produce radishes'. We therefore need to design an experiment where some plants receive adequate light (we have to decide how much light is adequate) and another set of identical plants receives low light levels (again, we need to work out what a 'low light level' is). Once we are sure we can provide normal and low light levels, the soil, temperature, water and fertilizer levels must stay the same for both groups of plants. We can hypothesise that 'If radish seedlings need adequate light, then reducing the light available to radish seedlings will reduce their growth rate'. This proposition says that reducing the light causes the plants to grow less. What we need to measure is clear (or is it?). What happens if the seedlings in low light grow taller but thinner than the seedlings with more light? (This often happens.) We may have to modify our hypothesis to ensure that we are measuring total growth not just height! Testing a hypothesis is not easy. The fruitfulness of a hypothesis depends on how it is stated and what predicted evidence supports it and what evidence will disprove it.

Bold hypotheses

Scientists like bold hypotheses—we can see exactly what they are proposing and it's relatively easy to design a test that disproves them (yes, we can disprove hypotheses and theories) or supports them (no, we cannot prove hypotheses or theories). If they can't disprove a popular theory, good scientists remember that this means simply that they have not yet thought of the clever or bold *experiment* that will test the theory. Good scientists are always sceptical, they recognise that their conclusions are provisional and could be disproved tomorrow, next month or in ten years' time. Peter Doherty (an earlier Australian Nobel laureate) says, 'Accept nothing at face value, and get in the habit of thinking unconventionally. Work hard, work smart and, with a bit of luck, serendipity will play its part' (2005, p. 16). Doherty's advice is interesting because he sees luck and serendipity playing a role in science. Or is he saying that if you propose enough hypotheses and work night and day doing enough intelligent experiments you'll strike gold? As Thomas Edison famously said, 'Genius is 99 per cent perspiration and 1 per cent inspiration'.

This last point helps refute the proposition that 'One good experiment could settle a scientific argument'. Successful experiments are repeated and changed by scientists who read about other scientists' work. This is standard practice. A scientist's work is rarely accepted until it is replicated successfully. In 1989 Stanley Pons and Martin Fleischmann claimed that they had achieved 'cold nuclear fusion'. It was a wonderful idea that promised limitless supplies of safe nuclear energy but the original 'successful' experiments could not be replicated. The theory was discarded; or was it? Ten years later US naval scientists in San Diego were quietly working on it. If 'cold fusion' theory ever does work, it will be only as a result of hundreds of experiments that extend the original work.

Is astrology science?

Questions 3, 8, 13 and 18 were included in the HITS-W survey because some people believe in extrasensory perception, water divining and astrology. Horoscopes are so general that they could apply to anyone and everyone and there is no fair test of astrology's claims for an individual or a group of people. Indeed, an AU$10 000 prize offered in the early 1980s by Dick Smith for accurate water divining (dowsing) has never been successfully claimed (see http://www.skeptics.com.au/articles/divining.htm). Scientists regard dowsing, spoon bending, dream interpretation and 'magic' as unscientific because they cannot be tested. Likewise, statement 19, 'Some scientific theories cannot be tested', is an incorrect statement. Since there is no reliable way of testing phenomena such as dowsing or spoon bending, all paranormal thoughts and actions must be treated as unscientific.

The history of science shows us that theories lead to experiments not the other way round. For instance, when astronomers discover a night-sky object they have never seen before, their first step is to check that no-one else has seen this object and studied it. If it seems to be unique, the astronomers will try to measure its size, brightness, shape and distance from the Earth. They will then propose what it is, using information contained in established theories to construct their hypothesis. They must do this because they cannot be sure the object is unique until they have compared it to every other known object in the night sky. The hypothesis will be an 'if–then' statement: *if* the object is x, and works in such-and-such a way, *then* it should do y. Such a prediction will be testable. Astronomers test new ideas by looking for other examples of the unique object. If they find other examples that agree with their predictions, their hypothesis is supported. Experiments that test a theory's predictions will do one of two things—support or disprove the theory. For tests and experiments to have meaning, they need to be based on theoretical predictions and they need to ask purposeful questions.

Predictions are crucial. A hypothesis that cannot make testable predictions must be discarded in favour of one that will. Each fruitful testing cycle starts with a hypothesis—if the hypothesis is credible, it will predict new actions, objects or behaviours. Each prediction must be unambiguous and most of the test results are anticipated. If the hypothesis is supported, such-and-such will happen; if the hypothesis is disproved, some predictable things will happen but there will be unexpected outcomes that will make you go back, modify the hypothesis and then retest it. The cycle goes on until either the hypothesis can be consistently supported or it needs to be discarded and a new hypothesis formed and tested. Hypotheses that make ambiguous predictions are always rejected.

Snapshot 1.3: Scientific theories that changed our world

Cell theory states that all living things are composed of cells or are the product of cells. Viruses and prions (microscopic protein particles which are similar to viruses but contain no nucleic acid and which cause fatal brain diseases such as mad cow disease and Kuru) are all made by cells and involve cells at some part of their 'life' cycle.

Evolution has been called the 'unifying' theory of biology because it explains the origins and relationships between all living things. Despite what we don't know about extinct plants and animals, scientists are fascinated by the usefulness of this theory.

Atomic theory grew out of the idea that matter could not be sub-divided beyond an imaginary limit (*atomos* can be translated as 'indivisible') and now describes atoms as sub-microscopic 'elastic balls', 'solar systems' or mathematical 'waves'.

Relativity says that nothing is absolute. According to Newton's Law of Gravitation, all objects are always attracted to all other objects. Newton's Law says that gravitational effects are instantaneous:

> that is, if we were to move a mass, we would at once feel a new force because of the new position of that mass; by such means we could send signals at infinite speed [across the universe] . . . but because we cannot send signals faster than the speed of light . . . the law of gravitation must be wrong. (Feynman, 1995, p. 112)

But Newton's Law of Gravitation still works perfectly for everyday objects and processes.

Quantum theory deals with the strangeness of the sub-microscopic world. In the natural world the distances animals, plants and rocks move, their weights and their energy changes are infinitely varied. In the world of electrons, protons and neutrons, however, everything is quantised—that is, it changes in fixed steps like steps on a ladder. Something either has a particular energy or it doesn't, there are no intermediate positions. Is this important? Yes. Because it explains how computers, quartz watches and solar cells work.

Plate tectonics is a modern theory that is widely accepted because it explains the origin of earthquakes and volcanoes and helps us predict disasters. Modern satellite images enable scientists to show that the Australian plate is drifting north at a rate of 2–3 cm per year.

Climate modelling isn't yet called a theory but it does most things you expect a theory to do. We now understand most of the details about ozone holes, greenhouse effect and sea-level change—the devil is in the detail but that's true for most theories.

SUMMARY

Scientific theories are built on robust, tested hypotheses. Evidence for theories accumulates as hypotheses survive repeated, rigorous tests. Repeating the same test many times will not improve our knowledge; each successful test must be new and different. Each additional piece of evidence gradually builds a theory by induction (Chalmers, 1999). No single piece of inductive evidence is conclusive on its own but each piece combines to increase our confidence in a theory. When a theory has stood the test of time, debate and many experiments, it is accepted as the best available explanation (see, for example, plate tectonics and cell theory). The theory can then be used to deduce what will happen in novel situations, and these deductions lead to new applications and tests. Deductions are much stronger and more reliable than inductions. The stronger a theory is, the broader and more far-reaching are its applications, predictions and uses.

Of course, every new application of a theory tests it and exposes it to falsification. But that's how science works. When a trusted theory fails, what usually happens is that the theory is changed. If the failure is critical (which is rare), then there is a revolution against the established order and new, competing theories vie for acceptance until one accumulates enough evidence to be accepted. Then the cycle starts all

Figure 1.1: A model of the science learning cycle

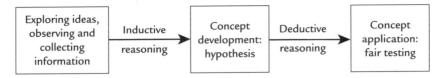

over again. When a revolution happens, several new theories exist side by side and that's good for science.

The learning cycle in Figure 1.1 models scientific thinking. The learning cycle means that it is possible for school scientific investigations to emulate some of the ways scientists 'do' science. In the learning cycle, students and teachers start by choosing a problem or problems where students can first explore what is known about the problem and then decide what they need to find out. This is the exploration phase and its primary aim is interest and excitement. Next, students use the information they know to propose a theory or hypothesis that can be tested. Theorising is important and the hypothesis should generate additional research questions and predictions that can be tested. If the students' hypothesis is too bold, the required tests may be beyond them; in this event, they should moderate the hypothesis until the test becomes doable. Scientists do this all the time and the teacher's role in the investigative cycle is to check on safety, suggest equipment and provide need-to-know information. Students often choose too small a sample or confuse their variables—both excellent opportunities for discussions that can help all the students in a class. In time, a series of predictions and tests should generate enough data that the students' hypothesis begins to resemble a concept. Once a reasonably stable concept emerges, the concept-hypothesis becomes deductive and 'if–then' reasoning and testing can be employed. If your first primary science investigations do not follow this script, do not be disheartened: the wonder of science lies in learning from mistakes and wondering why the unexpected happened. Science wouldn't be science and most of our great discoveries would never have happened if scientists always followed the script.

REFERENCES

Australians win Nobel for linking ulcers to gut bug. (2005, October 8). *New Scientist*, *2520*, p. 7

Chalmers, A. (1999). *What is this thing called science?*. St Lucia, Qld: University of Queensland Press.

Doherty, P. (2005, October 15). Soundbites. *New Scientist*, *2521*, p. 16.

Fensham, P. (2005). A school science curriculum. Unpublished paper presented at a *Science Works for the Smart State* workshop at Central Queensland University, Bundaberg, 10 August.

Feynman, R.P. (1995). *Six Easy Pieces*. Reading, MA: Helix Books.

Friedman, M., & Friedland, G. (1998). *Medicine's 10 greatest discoveries*. London: Yale NotaBene.

Harrison, A.G. (2005). *Understanding and teaching primary science*. Rockhampton, Qld: Central Queensland University.

Latour, B., & Woolgar, S. (1986). *Laboratory life: The construction of scientific facts*. Princeton, NJ: Princeton University Press.

Millar, R., & Osborne, J. (1998). *Beyond 2000*. London: Kings College.

Schwab, J.J. (1966). *Teaching of science: The teaching of science as enquiry*. Cambridge, MA: Harvard University Press.

Tobin, A.J., & Dusheck, J. (1998). *Asking about life*. New York: Saunders College Publishing.

CHAPTER 2
VIEWS OF STUDENT LEARNING

Coral Campbell and Russell Tytler
Deakin University, Victoria

OUTCOMES

By the end of this chapter you will:
- recognise the importance of students' prior conceptions in framing learning in science;
- be able to describe recent shifts in perspectives on student learning in science; and
- appreciate the broader affective and personal factors that shape students' learning in science.

INTRODUCTION

In this chapter we will trace the changes in views about student learning which have taken place over the last few decades. These changes are largely in response to a significant body of research into students' ideas about science, and the particular challenges involved in supporting meaningful learning of major science concepts. Many of the findings about students' conceptions (a 'conception' being a way of looking at a particular phenomenon) of everyday phenomena are surprising and intriguing, and now well-established. Interpretations of these findings have shifted, however, from a focus on learning as a change in individuals' conceptions, to sociocultural perspectives that place more emphasis on the social context of learning and acknowledge the role of language and culture. There is considerable debate about how best to think about the acquisition of science knowledge, and what we should focus on to support students' learning. Should we focus on challenging the conception held by individual students, or be more concerned with setting up rich and productive ways of talking and interacting in classrooms? There has also been considerable research in recent years on other aspects of children's learning in science, including reasoning, motivational factors such as goals, values and interests, children's view of themselves as learners and creative thinking.

STUDENTS' SCIENCE CONCEPTIONS

There is a large body of research on the conceptions students bring with them to the science classroom, and how these affect how and what they learn (see, for example, Duit, 2002). This interest in students' conceptions stems from the realisation that children can emerge from a science learning sequence with understandings very different to those intended by the teacher. Even students who perform at a high level on classroom tests may display a range of very different understandings when asked to apply these ideas to other situations, especially out-of-school contexts. The main findings of this body of research can be summarised thus:

- students come into our classes with a range of prior ideas or conceptions of the physical world. They are not 'blank slates' waiting to be written on by the teacher;

- many of these conceptions are commonly held, and differ in important ways from the view of the world scientists have constructed; and
- these conceptions in many cases form useful prior knowledge that a teacher can build on. In other cases, however, students' alternative conceptions have proved surprisingly difficult to shift, and can be a serious barrier to learning.

The following are some of these alternative conceptions that interfere with students' learning across a range of science topics.

- Students believe that plants receive their food through their roots and find it inconceivable that, through the process of photosynthesis, plant material originates in atmospheric gases and water. This idea has obvious, common-sense associations with human ingestion and is reinforced by everyday language (e.g. 'plant food').
- Young children interpret animal behaviour in terms of the animal's wishes (a psychological view), and find it difficult to think of behaviour as having an adaptive function. Thus, a Year 3/4 child will say, when asked why worms congregate in moist soil when given a choice: 'Because they might be dry and they wanted to be wet'. When asked about the effects of introducing European carp into Australian waterways, Year 3/4 children tend to think in terms of direct effects such as predation or muddying the water, whereas Year 5/6 children are more likely to talk about the interaction effect of competition for food. Similarly, in response to the question 'Why do plants have flowers?' the main responses selected by Year 2 children are 'to make them look nice' or 'because bees need the pollen and nectar', whereas older children look for explanations involving adaptation.
- Students have a 'historical' view of substances in chemical change. They think, for instance, that the ash left over from burning paper is either the paper but in a changed form, or something that was trapped in the paper and is now the residue; or that rust is simply iron in a different, broken-down form.
- There are a number of alternative conceptions arising from the perceptual difficulties associated with air. For instance, primary school children will be amazed that a tissue in an upturned glass remains dry when plunged into water. The tendency to cast explanations in

terms of things we can perceive directly also underlies the difficulty students have with explaining the condensation that forms on the outside of cold objects, since they have no idea that air contains water in gaseous form.

- Students believe that heat is a substance, rather than a form of energy, and run the concepts of temperature and heat together. This causes them to think, for instance, that if a hot cup of coffee is divided, the temperature is halved.

Significant alternative conceptions are often associated with concepts for which the everyday and scientific use of language differs—'animal', 'energy', 'force', 'heat' or 'alive', for example. Such conceptions can be extremely difficult to shift, supported as they are by everyday language. Many studies (Bell, 1993; Venville, 2004) have demonstrated how alternative conceptions can persist despite carefully planned teaching, and often despite students being able to gain high scores on science tests.

How should we view these conceptions?

The different terms used for these conceptions reflect very different ways of viewing them. The term 'misconception', which tends now to be out of favour, implies that these ideas are simply 'wrong' when judged against 'correct' scientific conceptions. Such an interpretation devalues the very real conceptual content of some of these ideas, and implies teaching approaches aimed at eradicating misconceptions. This in turn tends to encourage transmissive teaching strategies. The term 'children's science', on the other hand, promotes a view of children as fledgling scientists constructing, from their everyday experience, conceptions that have some of the characteristics of scientific theories (a certain degree of consistency and explanatory power, for instance) but are perhaps limited in scope of application. A similar emphasis is implied by the term 'alternative frameworks'. If this is the case, then science teachers must find ways to provide pathways, or 'bridges', that will support students as they undergo major shifts in their perspective towards a scientific way of looking at the world.

This 'conceptual change' idea received a lot of support from researchers through the 1980s and 1990s (Hewson & Thorley, 1989; Duit & Treagust, 1998). The idea is appealing partly because of the intriguing similarities between children's ideas and those of earlier scientists, and partly because it links with the emergence of a view that major scientific advances involve revolutionary 'paradigm shifts'. Thus the change from an Earth- to a sun-centred solar system, or the emergence of a continental drift theory, are regarded as revolutions involving quite new ways of looking at the world. The argument is that students' conceptual change involves similar, personal revolutions in their perspectives and, as in science, these 'revolutions' may happen over a number of years. Children move from the psychological view of animal behaviour, described above, to a biological view of behaviour as adaptive slowly over their primary school years.

Why should this issue matter for us as teachers? If we consider that teaching and learning science often involves a process of change in fundamental perspective, then it explains why learning about animal adaptation, or chemical change, can be so difficult. Science ideas can be quite contradictory to students' prior perspectives and this means that we cannot consider teaching as simply the clear explanation and demonstration of new ideas. We need to recognise students' prior views, and be strategic in challenging them. On the other hand, it is helpful for us to realise that teaching young children about adaptation, for instance, is a long-term process that involves gradually chipping away at their naïve views, and we should not be disheartened by incomplete adoption of the biological view.

Should we focus on eradicating these views? A number of writers (e.g. Inagaki & Hatano, 2002) have argued that anthropomorphism (thinking that animals and plants make decisions or behave like humans), for instance, can be useful as a springboard for developing more sophisticated scientific ideas. For example, if a student initially thinks that 'plants have flowers to look nice', this idea can be used as a springboard to develop the idea that flowers are often attractive to pollinating insects and birds. Other writers (Lautrey & Mazens, 2004) have questioned whether students are consistent in the way they use either alternative, or science ideas. There is evidence (Venville, 2004) to suggest that students do not abandon their intuitive conceptions once they learn to operate with a science conception, but keep them in the background, to be applied in particular situations. Even adult scientists will carry a range of views about animals, or energy, or 'suction' alongside their more sophisticated understandings. Recently, there has been acceptance of the view (Liu & Lesniak, 2006) that understanding and learning are richer and more complex than the simple conceptual change model implies. Thus, as teachers we should recognise the richness and variability of students' ideas, and find ways to both challenge and make use of them.

CONSTRUCTIVIST PERSPECTIVES ON LEARNING

Research into children's conceptions has raised questions about the nature of student learning, and about what we might mean by 'understanding'. From the start, researchers Duit and Treagust (2003) recognised that learning should not be seen as some sort of conceptual implanting process (the 'jug and mug' metaphor!), but as an interplay between students' existing ideas and the knowledge or experiences they are exposed to in the classroom. This perspective views learning as the construction of personal meaning, and learning in the classroom becomes an extension of the process by which the prior ideas were developed; from students' active engagement and meaning-making with the world around them. This personal, conceptual view of learning owed much to the child development theories of Piaget and has become known as the 'constructivist theory of learning' (Duit & Treagust, 1998). In a personal constructivist view:

- learning involves the active construction of meaning from experience;
- meanings constructed by students from what they see, hear or otherwise experience in class may be different to those intended, and are influenced by their prior knowledge, which can assist but can also interfere with learning; and
- learners have the final responsibility for their own learning. It is the teachers' role to promote opportunities and support for learning. This does not rule out, of course, the effectiveness of clear explanations and structured discussion, but it does emphasise the essentially personal nature of students' pathways to understanding.

Notice that, while this view acknowledges the importance of language (a social phenomenon) in the construction of meaning, the emphasis is very much on the individual's construction of conceptual meaning.

That is why we are calling it a 'personal constructivist' view. In science education, constructivism's appeal grew out of the need to interpret the empirical findings of student conceptions research. However, constructivism has become well-established in the literature across a range of educational areas. It is a 'broad church' containing many sects (Phillips, 1995).

We must emphasise that constructivism is a theory of learning and does not necessarily imply a teaching program. It is possible, for instance, to talk of students constructing meaning from lectures or explanatory passages from books. Nevertheless, constructivism does have some well-established implications for teaching, including the need to explore students' prior knowledge, to monitor students' understandings and to support students' active engagement with science ideas. A number of 'conceptual change' approaches to teaching have been described and researched, all of which are based on the idea that learning involves a major restructuring of conceptions rather than the presentation and exploration of established scientific knowledge. These approaches (for example, the 5Es approach described in Chapter 6) involve exploring and challenging students' prior ideas, establishing the science ideas, extending these ideas to a range of phenomena and explicit evaluation of the new perspective. Conceptual change approaches have been shown to lead to significant learning, with a major factor being the explicit challenge to prior ideas, and support for the restructuring process.

One of the difficulties in talking and making judgments about constructivist-inspired teaching is that its ideas tend to draw on a long history of ideas in science education. Thus, traditions of 'inquiry' and 'hands-on' teaching are incorporated into responses to constructivist ideas. The older idea of 'discovery learning', that students should learn for themselves using structured activities, and that teachers should never 'tell' but only guide and question, is an extreme interpretation of constructivist ideas that has led to constructivism gaining a dubious reputation in some educational circles. Mainstream conceptual change approaches have raised questions about the time-consuming nature of constructivist-inspired strategies for both teachers and students. Indeed, constructivist writers vary considerably on the time they recommend teachers to spend in exploring and negotiating students' understandings, rather than directly representing the science view.

SHIFTS TOWARDS A SOCIAL CONSTRUCTIVIST PERSPECTIVE

Within the constructivist perspective itself, the ground has been shifting. There is a substantial body of criticism of both the early constructivist literature and the conceptual change literature, pointing out the narrowness of such a purely conceptual view of learning and its excessive focus on the learner at the expense of the teacher and classroom strategies (see, for example, Duit & Treagust, 1998). Since the 1990s there has been increasing interest in exploring learning as a 'social' or cultural phenomenon, with a shift in focus from individual students' understandings to the ways in which classroom environments can support effective learning.

A social (sometimes called sociocultural) constructivist position focuses on the social processes operating in the classroom by which teachers promote a discourse community in which they and their students 'co-construct' knowledge. The aim of science education then becomes the establishment within the class of shared meanings (Driver, Asoko, Leach, Mortimer & Scott, 1994). In this process the teacher represents the very powerful language and practices of scientific culture, and scientific ways of viewing and dealing with the world. These ideas are strongly associated with the writings of a Russian psychologist, Lev Vygotsky (1986), who emphasised the role of language and culture in framing the way children learn to interpret the world, rather than placing most attention on the individual grappling with their experience.

Some see social constructivist perspectives as putting the teacher back in the picture, in contrast to the personal constructivist position, which many felt was writing the teacher out of the learning process. A social constructivist view focuses on the teacher interacting with their class, whereas a purely personal constructivist view focuses on what is happening in individual students' minds. A personal constructivist view suggested personalised learning programs based on probing students' prior conceptions—a very difficult project for a teacher.

As an example of an approach that focuses on the setting up of a community of inquiry in the classroom, Suzanne Peterson (Tytler, Peterson & Radford, 2004) designed an Animal Behaviour unit for her Year 3–4 class which focused explicitly on clarifying the nature of science processes such as observation and inference. Students created lists of observations, and the inferences they might draw from each. The unit also focused on supporting children to develop the language of science reasoning and evidence by asking questions such as 'How can you show that?' or 'How could we find out?' Peterson designed guided explorations on the behaviour of selected animals (crickets, fish, birds) in which measurement processes (such as 'How to track movement') were modelled, as well as experimental designs based around students' questions and predictions. While the focus was strongly conceptual in that the children were exploring ideas about animal behaviour and adaptation, more emphasis was placed on establishing a high level of shared language around investigating ideas and evidence than on explicitly probing individual conceptual positions. When these surfaced, they were negotiated as part of the discussion.

A complementary perspective is provided by Scott (1998), who argues that the key to understanding effective science teaching and learning is 'classroom discourse' or the pattern of teacher-and-student talk during science lessons. Scott argues that in effective lessons, classroom discussion switches between sequences that are open and exploratory and in which students throw in lots of ideas (dialogic discourse), to sequences where the teacher exerts more control over the conversation to bring these ideas together (authoritative discourse). The ability to shift between these different discourse modes is an important aspect of effective teaching. Recent research has explored this idea.

Corresponding to these social constructivist perspectives is the growing interest in theories of learning that give a more fundamental role to language and culture in the construction of knowledge and in the way we think. Vygotsky's ideas provide a framework to argue for the central role of discourses and communities in knowledge production. This 'sociocultural' perspective can be understood to correspond to the 'social constructivism' used sometimes in science education, but tends to portray knowledge as a property of participation in a discourse community, rather than something acquired by individuals through social processes. It provides an opposing perspective to personal constructivist and conceptual change views and sees learning not in terms of the acquisition of stable, resolved 'conceptions' or 'mental models' by which the mind represents the world, but as increasing students' access to and competence with ways of talking and acting within the discourse community.

A further perspective that emphasises social contexts of learning is the *situated cognition* perspective, which overlaps considerably with sociocultural theory. Rogoff and Lave (1984) argue that context is an integral aspect of what we learn; the situation in which an understanding is generated is an integral part of that understanding. In this view, learning moves from being a peripheral to a central participant in a discourse community as students achieve increasing competence in the specific ways of talking and acting in that community. Learning science, therefore, is about participating in scientific community practices, and in this sense the school science community should aim to represent authentic science practices. Science educators have long recognised that knowledge is a product of the context in which it is learned and used. Joan Solomon (1983) described the way in which real-world contexts triggered different views and explanations of energy, compared to when questions were asked in a classroom context. She was able to show that, over time, the learned scientists' view of energy was abandoned in favour of 'life-world' understandings, as these everyday forms of knowledge reasserted themselves outside the influence of the science classroom. The implication of this insight must be that if school knowledge is to be useful to learners out there in the world, continual links must be made between school experience and social uses of science knowledge. We must break down the boundaries between the classroom and the rest of the world.

Learning and literacy

So is learning a personal, or a social, phenomenon? Both perspectives, the individual and the social, offer powerful insights into students' learning, and it would be foolish to abandon either.

A large part of the critique of the individual perspective concerns the characterisation of learners' conceptions as things that will be fixed in people's heads once an understanding is achieved. A blending of the individual and the social perspectives requires us to develop more subtle views of the way in which individuals adopt the language practices of science in coming to deeper understandings. There is growing interest in the notion that 'understanding' something inevitably involves representing it in some way, either through written or spoken language, gesture, diagrams, or 3-D models, and that the process of learning involves re-representing ideas in these different

Figure 2.1: Representations of evaporated alcohol in the room. Karen's understanding of evaporation was significantly advanced through the negotiation of her representation of particles with an adult.

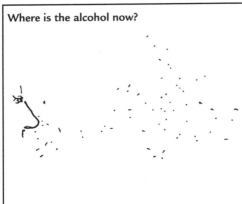

Where is the alcohol now?

Verbal account: The little molecules would be floating upwards. (I: How are you able to smell that?) Because there are so many of those little molecules passing up and even though it is such a small drop there would be so many that they would go past my nose and would actually go up my nose. (I: So you are just drawing a few around your nose now?) They will be everywhere.

modes. The representational resources that children need to develop are the literacies of science.

We (Tytler, Peterson & Prain, 2006) worked with Year 5 children, supporting them to represent the process of evaporation using a molecular model introduced with plastic beads. We found that exploring the process of evaporation of an alcohol drop with individual children—using verbal explanations of a drawing (visual representation) of the evaporated alcohol in the room—led to a powerful revision of ideas and learning (see Figure 2.1). Using drawings to represent the idea of particle distribution and evaporation, and coordinating these drawings with written or spoken language, as well as gestures or 3-D models, is the Vygotskian process of transporting language from the social group of learners to the individual minds. (See Chapter 10 for a further discussion of the literacies of science, which are built around the idea of representational resources.)

HIGHER-ORDER ASPECTS OF STUDENT LEARNING IN SCIENCE

If we view learning as increasing students' capacity to participate in the literacies of science, then this will involve much more than conceptual understanding. Driver, Leach, Millar and Scott (1996) argue for the importance of students' reasoning strategies, and their views on knowledge production and the nature of science. Current perspectives on higher-order thinking, essential learnings and the 'thinking curriculum' emphasise such things as reasoning, metacognition (thinking about and managing our own thinking), imagination and creativity, and individual perspectives on the learning process. Other writers (e.g. Sinatra, 2005) have also emphasised the importance of affective factors such as motivation, interest and values. Thus, there is a growing body of opinion that in order to understand the nature of science learning in classrooms, and provide advice for the effective teaching of science, teachers must look beyond the individual conceptual focus, and take a wider view of the learner and the process of learning.

Learning to reason in science

A major goal for school science is that students learn to *investigate* science, which involves reasoning processes such as questioning, designing experiments and measurement methods, controlling variables, and analysing evidence to support arguments. These processes are described further in Chapter 7. Reasoning, however, is not confined to practical investigation. Learning and applying the concepts of science both involve reasoning, and there is evidence that knowledge and reasoning are interdependent. Some writers, following Piaget, think of reasoning as a general ability that is determined by a developmental sequence. It is often measured in terms of how well students can control variables in decontextualised situations. There is mounting evidence, however, that young children are capable of significant reasoning, and that children's reasoning capacity is best demonstrated in contexts familiar to them.

Reasoning involves being able to link science ideas with evidence; thinking of ways of testing ideas with observations or experiments; and using observations and evidence from experiments either to justify ideas, mount arguments, or think of alternative explanations. Part of reasoning is learning how to explore and generate ideas. We (Tytler & Peterson, 2005) found that in hands-on explorations young children operated at a range of levels, from randomly trying things out and generating unrelated explanations, through looking for patterns, to the highest level where they developed an idea and tested it by targeted exploration. As an example of the highest level, which is the essence of scientific reasoning, Jeremy (a Year 2 child) was exploring the factors that affected the flight of whirlybirds (small 'spinners' made of folded paper weighted with a paper clip). There is a definite sense here of an active program of trying out ideas:

> I'll try it . . . yes, slower . . . the paper clip puts weight on it to make it go a bit straight. And also . . . if the wings are really long then they might work a bit more because it will have more air in it. The air . . . is going up under the wings and then when it's going down it makes it spin more. I'll get another one. Yeah, I think this one will take a bit more time to get down . . . because when the wings go in there, it's like a parachute coming down and when a parachute opens it lets you go down you get slower . . . and so does that [he tests and his idea is confirmed].

We found that children's ability to reason depended on their science knowledge—no-one can reason in a vacuum, and Jeremy needed a concept of air, and of gravity, and a knowledge of parachutes to reason as he did. Conversely, reasoning is an integral part of learning to establish and work with ideas. Thus, explorations of the type illustrated in Jeremy's transcript provide powerful learning opportunities.

Reasoning is also related to effective learning strategies such as monitoring and controlling learning, or asking questions, which have been the focus of the influential Project for the Enhancement of Effective Learning (2005), which pioneered ideas about the 'thinking curriculum' and metacognition as a key to effective learning.

Creativity

Einstein has been quoted as saying that imagination is more important than knowledge. What is the relationship between children's learning and creativity and imagination? Definitions of 'creativity' suggest that it is the ability to produce novel work or ideas that are appropriate to the constraints of the task. Stated more simply, creativity is about making connections where none existed previously.

The recent introduction of the 'thinking curriculum' in classrooms has concentrated on developing children's higher-order thinking skills—synthesis (creative thinking), analytical thinking and evaluation. Many recent curriculum documents promote and encourage higher-order thinking, student-centred learning, student independence and self-motivation. Student learning is said to be enhanced by the use of thinking strategies such as De Bono's 'Six Thinking Hats' (De Bono, 1985). A classroom structure that successfully incorporates higher-order thinking will employ a careful balance between content knowledge and process.

Tackling a science task may require students to think divergently, generating new ideas (creative thinking). They may also be required to think convergently (analytical thinking) and to evaluate appropriate solutions. What specific circumstances support either creative or analytical thinking at a particular time, and how does this impact on students in the classroom?

It is assumed that problem solving in science education engages children in creative thinking, and that creative problem solving can be enhanced through teaching, particularly where instruction and counselling are given. To foster creativity, children must have some background knowledge and experience of the relevant area, be able to think divergently as well as convergently, have the capacity to analyse and synthesise the problem, and be able to enact self-planning and evaluation strategies. Creative thinking is improved when students take part in an interactive learning process and trial different strategies. Teachers can improve students' creativity further by using specific contexts to model the different problem-solving processes (for example, building a wind-up vehicle). This modelling process enables students to transfer the learning skills to new situations (for example, building a wind-powered boat).

THE CHILD AS A LEARNER—AFFECTIVE FACTORS

Children's view of themselves as learners of science (Am I interested in science? Am I 'good' at science? Do I like to explore things in science?) frames their approach to any science learning task. Children vary markedly in this regard from the first year of school. For instance Calum, three months into the first year of school, described doing science at home.

> Well in our science book we tried growing crystals but it took a long time but it didn't actually grow. We needed baking powder and water

and a bowl of water and a jar of water and a rope and a pencil . . . and we tied the pencil to the rope, the string, and then we put the pencil in, we lowered it down and the pencil it was balancing where the lid would go on. And we put . . . the paperclip didn't actually get crystals on it only the rope.

Calum had a very different perspective on himself as a science learner compared to some other children who had a view of learning as the passive acquiring of factual knowledge. He retained this view of science learning as active and inquiry-based throughout his primary school years.

Children's approaches to learning in science are strongly influenced by how they interpret the nature of a task and its context, and the way they characterise themselves as learners and relate to the task. Classroom activities can be very social and conceptually-rich events, and children may focus on different aspects of a task. For instance, Figure 2.2 shows Sophie's (Preparatory Year) drawing of a first-year evaporation activity involving tracing the shrinking of a puddle in the playground. Her explanation was dictated to her teacher, but the drawing clearly illustrates her focus on the community aspect of the event. Sophie, in interview, reflected a strong social narrative of her place in this task, alongside her narrative of herself as a science thinker.

Much of the research in science education in the last 20 years has been focused on the cognitive aspects of science learning. Most of us are aware, however, that students who are not happy and stable in their classroom will not learn. Learning is not only supported but enhanced through positive affective factors. These factors are recognised as relating to student interest, motivation, attitudes, beliefs, self-confidence and self-efficacy. The role of the social context is extremely important in establishing these factors through students' relationships with teachers, peers and other important connections.

Motivation

What qualities come to mind when we describe motivation? Motivation, student beliefs, interests and goal orientation are key factors in student learning. Students' beliefs in their own abilities (self-efficacy) and the value they attribute to their learning tasks are significant

Figure 2.2: Sophie's drawing of an evaporating puddle activity. The black dots represent the tops of students' heads.

When we left the puddle might have disappeared. It went into the sun.

predictors of their final success. In addition, motivational beliefs can influence the process of learning and conceptual change. This motivation is usually context specific and relies strongly on the classroom situation. Therefore the sociocultural context becomes an important determinant of learning.

Students need to believe that they are capable of performing a task, that they have some control over the task and that the task is achievable. These factors will vary depending on the classroom context, the chosen task and the effectiveness of the teaching strategy. The level of engagement and willingness to persist in a given task can be attributed to a student's motivation. Students choose whether or not to become engaged cognitively with the task at hand. If they become engaged and persist, then learning is likely to occur. If they do not become sufficiently involved in the learning task, then the best result is likely to be surface learning.

Goal orientation and engagement

When undertaking any science task, a student's goal will guide how he or she engages with, and completes, the task. Those students who focus on obtaining the best result (competitive) as distinct from achieving mastery over the task (commitment to learning), have different goals and different levels of cognition. Students who engage with the task are more likely to do so at a deeper level of understanding whereas students with a competitive goal achieve more surface learning. The deeper the level of engagement, the higher the probability that conceptual change will occur. The nature of the task also influences the student's goal orientation—tasks that are meaningful, challenging and related to life outside school provide strong motivational goals. Open-ended classroom activities (see Chapter 6) are also more likely to facilitate cognitive involvement and conceptual change (Pintrich, Marx & Boyle, 1993). Where students are offered choice and some control over their learning, they are also likely to be more goal-oriented and hence more likely to learn.

Values and interests

Other aspects of motivation are a student's value of, and interest in, science. For example, a student may be extremely interested in an area of study, but may also value it in terms of its importance in other aspects of life. These value and interest beliefs are said to be intrinsic, that is, they are characteristic of the person, not the task. Some students really like science, but some are less interested. A student's personal interest will influence their selective attention, their effort levels, their willingness to persist with a task and their final acquisition of knowledge. Students who are interested in class material seek further information, engage in more critical thinking and demonstrate elaboration strategies (those practical aspects a student will use to find out more about a topic). Further work has shown that when contexts are chosen for their high interest or value to students, the students demonstrate more competent learning through persisting longer at the task. There is strong evidence (Sinatra & Pintrich, 2003) to suggest that the value a student places on a task or topic will affect their choice to become engaged, support their learning by increasing attention and persistence, and activate prior knowledge. At a different level, a student's interest can be influenced by motivational features of the classroom context such as challenge, choice, novelty, fantasy and surprise. All of these aspects need to be kept within the capabilities of the students.

Beliefs-in-self

Another important aspect of motivation is a student's belief that he or she is capable of completing a task in a manner satisfactory to them. This links very strongly with the previous discussion on 'the child as learner' and a child's identity within the science classroom. These beliefs-in-self (self-efficacy) are said to be situation-specific, not general self-concepts (Pintrich et al., 1993). In science, this means that students need to feel comfortable and confident in their understandings and capabilities. As classroom conceptual change strategies tend to challenge students' accepted understandings, they may also shake a student's confidence in him or herself. Students with low self-confidence with regard to science learning and understanding can feel threatened by the conflict of conceptual challenges, although this also means that they are less likely to retain their own prior conceptions. Conversely, students with high self-esteem, and more belief in their own judgments, are more likely to retain their own developed conceptions. However, studies have indicated that students with strong confidence in their own abilities will demonstrate conceptual change when challenged through persistent and aggressive interaction with materials and ideas.

SUMMARY

In this chapter, we have highlighted recent research into children's learning in science. Constructivist theories have been expanded from an individualistic framework to include the sociocultural nature of learning and knowledge. The conceptual change perspective on learning has been expanded to include the idea of learning communities and learning as participation. We then took the discussion beyond these purely conceptual perspectives to consider children's reasoning, affective factors such as motivation and beliefs, and children's identity and creativity, as new and fresh perspectives on learning.

It is very clear that learning involves a complex set of physical, cognitive and behavioural factors, all of which must be considered in enhancing students' learning opportunities. Teachers need to be aware not only of the conceptual theories of learning, but wider perspectives that include multiple aspects of children's thinking, their view of themselves and their commitments and interests.

REFERENCES

Bell, B. (1993). *Children's science, constructivism and learning in science.* Geelong: Deakin University Press.

De Bono, E. (1985). *Six Thinking Hats.* Toronto: Key Porter Books.

Driver, R., Asoko, H., Leach, J., Mortimer, E., & Scott, P. (1994). Constructing scientific knowledge in the classroom. *Educational Researcher, 23*(7), 5–12.

Driver, R., Leach, J., Millar, R., & Scott, P. (1996). *Young people's images of science.* Buckingham, UK: Open University Press.

Duit, R. (2002). Bibliography—Students' and teachers' conceptions and science education. Retrieved August 28, 2006, from http://www.ipn.uni-kiel.de/aktuell/stcse/stcse.html.

Duit, R., & Treagust, D. (1998). Learning in science—From behaviourism towards social constructivism and beyond. In B. Fraser & K. Tobin (Eds.), *International Handbook of Science Education* (pp. 3–25). Dodrecht: Kluwer.

Duit, R., & Treagust, D.F. (2003). Conceptual change: A powerful framework for improving science teaching and learning. *International Journal of Science Education, 25*(6), 671–688.

Hewson, P.W., & Thorley, N.R. (1989). The conditions of conceptual change in the classroom. *International Journal of Science Education, 11*(special issue), 541–553.

Inagaki, K., & Hatano, G. (2002). *Young children's naïve thinking about the biological world.* New York: Psychology Press.

Lautrey, J., & Mazens, K. (2004). Is children's naïve knowledge consistent? A comparison of the concepts of sound and heat. *Learning and Instruction, 14*(4), 399–424.

Liu, X., & Lesniak, K. (2006). Progression in children's understanding of the matter concept from elementary to high school. *Journal of Research in Science Teaching, 43*(3), 320–347.

Phillips, D. (1995). The good, the bad and the ugly: The many faces of constructivism. *Educational Researcher, 24*(7), 5–12.

Pintrich, P., Marx, R., & Boyle, R. (1993). Beyond cold conceptual change: The role of motivational beliefs and classroom contextual factors in the process of conceptual change. *Review of Educational Research, 63*(2), 167–199.

Project for the Enhancement of Effective Learning. (2005). Retrieved August 28, 2006, from http://www.peelweb.org.

Rogoff, B., & Lave, J. (Eds.). (1984). *Everyday cognition: Its development in social context.* Cambridge, MA: Harvard University Press.

Scott, P. (1998). Teacher talk and meaning making in science classrooms: A Vygotskian analysis and review. *Studies in Science Education, 32*, 45–80.

Sinatra, G.M. (2005). The 'warming trend' in conceptual change research. The legacy of Paul R. Pintrich. *Educational Psychologist, 40*(2), 107–116.

Sinatra, G.M., & Pintrich, P.R. (Eds.). (2003). *Intentional conceptual change.* Mahwah, NJ: Lawrence Erlbaum.

Solomon, J. (1983). Learning about energy: How pupils think in two domains. *European Journal of Science Education, 5*(1), 49–59.

Tytler, R., & Peterson, S. (2005). A longitudinal study of children's developing knowledge and reasoning in science. *Research in Science Education* (Special edition on longitudinal studies of student learning in science), *35*(1), 63–98.

Tytler, R., Peterson, S., & Prain, V. (2006). Picturing evaporation: Learning science literacy through a particle representation. *Teaching Science, 52*(1), 12–17.

Tytler, R., Peterson, S., & Radford, T. (2004). Living things and environments. In K. Skamp (Ed.), *Teaching primary science constructively* (2nd ed., pp. 247–294). Melbourne: Thomson.

Venville, G. (2004). Young children learning about living things: A case study of conceptual change from ontological and social perspectives. *Journal of Research in Science Teaching, 41*(5), 449–480.

Vygotsky, L.S. (1986). *Thought and language.* Cambridge, MA: MIT Press.

CHAPTER 3
UNDERSTANDING HOW CHILDREN LEARN SCIENCE

Ruth Hickey
James Cook University, Queensland

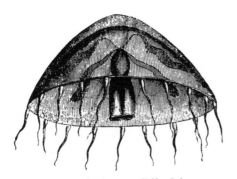

A Sea-Blubber or Jelly-fish.

OUTCOMES

By the end of this chapter, you will be able to:
- analyse children's responses to identify their science concepts;
- track children's conceptual development from early ideas through an increasing complexity of understanding;
- apply key ideas of science teaching and learning to practical situations; and
- identify teaching strategies which support conceptual development.

INTRODUCTION

This chapter will take you through four key ideas about how children learn science, with particular attention to how their concepts develop. These key ideas are illustrated by topics which are included in many primary science curricula: the phases of the moon, heat energy, properties of materials and the seasons. For each topic, you will complete an activity and reflect on your own science concepts before asking children of different ages to do the same activity. By comparing their responses, you can identify what happens as children learn science. This skill is critical when planning effective programs which support children to develop their science concepts. An analysis for each activity is provided, and implications for effective, practical teaching are suggested.

A common view of science is that it is 'all about facts' which are 'true' that people 'know' or can 'get right' in a test, for example, *Dr Karl's Collection of Great Australian Facts and Firsts* (Kruszelnicki, 2002). This view suggests that the purpose for teaching science is for children to learn that 'light travels at 300 000 km per second' or 'a tree is a type of plant'. While there are scientific facts we should learn, teachers can engage children in science successfully if they view science learning as 'getting the full story' not just 'getting it right'.

What do we mean by 'getting the full story'? Children develop early ideas about their world through their experiences. They build on these experiences (by observing, experimenting, listening or reading) in different contexts to develop consistent ideas, which enable them to predict events and feel confident about how the world works. They now have 'part of the story'. When children's experiences do not fit with what they already know, they revise their ideas. Through ongoing revision, children develop sophisticated concepts which are supported by extensive experiences, hypothetical and abstract thinking, and extended by ideas from books and media sources. Science becomes 'technical' when children apply mathematical thinking, multiple perspectives and complex understandings from multiple science disciplines to engage with an issue or problem. They are moving towards 'getting the full story'.

Rather than planning around 'science facts', teachers can plan for 'science concepts'. A science concept is an idea which we use to think about our world. For example, energy, light, gravity, the moon, plants, animals and respiration are all concepts. As children mature, they build up connections within concepts (for example, 'energy' includes its many forms such as potential, heat, light, sound, atomic, electromagnetic and kinetic energies) and between concepts (for example, an issues-based topic such as habitat preservation must consider energy webs).

What is a teacher's role when planning for conceptual development? Teachers should provide an environment which will foster children's engagement with science concepts, and support them to use and develop increasingly sophisticated concepts. This may be through providing many opportunities for children to develop their concepts, such as exploratory play. An ideal environment also encourages children to apply concepts to practical, interesting problems or issues, as in the 5E model approach of *Primary Investigations* (Australian Academy of Science, 1994) and *Primary Connections* (Australian Academy of Science, 2005). Ross, Lakin and Callaghan (2000) suggest the teacher's role is to 'use active learning approaches that allow pupils to reconstruct the scientific ideas we [teachers] present to enable them to make them their own' and if teachers want children 'to understand and use scientific ideas their existing beliefs need to be challenged or extended' (p. 29). Hand and Prain (1995) suggest that teachers need to rethink their teaching 'in order to have students more likely to reconsider the ideas and beliefs they [the students] already hold' (p. 7). It is also critical for teachers to be skilled in identifying children's concepts in order to use these judgments to inform further learning activities and guide assessment.

Think about how your own scientific knowledge developed. You may recall some of these developments in your own concepts, when you met ideas which were counterintuitive (not what you expected), and realised that:

- mushrooms, which you like, are actually a fungus;
- moonlight is really sunlight;
- sharks get new teeth for their whole lives;
- your belly button is a scar;
- fish breathe underwater;
- an iron will sink in water but float on mercury;
- squishy, green caterpillars turn into beautiful butterflies;
- cloning is a natural event (for example, identical twins); and
- you can't fly no matter how fast you flap your arms.

Similarly, new scientific evidence may be sufficiently challenging to cause the science community to re-examine long-held views, as part of the continual process of 'getting the full story'. For example, dinosaurs traditionally were portrayed as cold-blooded, slow and not too bright, but recent reinterpretations of fossil evidence suggest that many of them were fast, smart and socially cohesive, as they are depicted in the *Jurassic Park* movies.

Below are four key ideas about conceptual development, each with a practical activity. Each activity asks you to interview children from a range of ages, so you can work with children from four to 16 years old, and even adults. Take notes or tape each interview (with the children's and parents' permission) to help review and compare responses. Take time to do the activity yourself, and keep a journal of your reflections about your own conceptual development.

DEVELOPING CONCEPTS

Key idea: Children's conceptual development is characterised by increasing sophistication or complexity.

The moon and the solar system are commonly studied in the science topic Earth and Beyond. While you would expect everyone to know something about the moon, during this activity you will find that children and adults hold different ideas about its movement.

Activity 3.1: Can you see the moon during the day?

Write your own answer to this question and include an explanation. Then, ask some children and adults. Find some who say 'Yes' and ask them to explain their reasoning. Find some who say 'No, it only shines at night' and ask for an explanation. Listen to their responses, and look for differences in how complex their responses seem.

Children will vary in their beliefs and explanations about whether it's possible to see the moon during the day. Some have learned to associate moonlight only with night time (for example, many pyjama prints show the moon and stars).

Snapshot 3.1: Conceptual development of the concept 'moon'

Early concepts (ages 2–4+)

Children learn to point at something called 'moon'; know that sometimes there is moonlight at night; and may repeat inaccurate sayings such as 'the moon is made of cheese'. (Children's entertainment can reinforce these misconceptions, for example, Wallace and Gromit visit the moon and it *is* made of cheese.)

Consistent views (ages 4–6+)

Children may find the moon unremarkable, because they have sufficient amount of 'the story' not to have their views challenged: they can tell you it shines at night, is sometimes big and sometimes small.

Multiple revisions (ages 6–10+)

In response to learning activities in primary school, children may revise and expand their ideas as they incorporate new ideas: that the moon is a huge ball of rock (not cheese) that is going around the Earth; that it is hundreds of kilometres away; it is not a planet and not a sun; and that astronauts have landed on the moon.

Sophisticated applications (ages 10–12+)

Children develop new concepts: they plausibly link the phases of the moon to the Earth's tides; they accept that moonlight is reflected sunlight; they show how the moon orbits the Earth in a lunar month (28 days); they state that on the moon you can jump higher because of lower gravity; and grasp some of the considerations of space travel (for example, take your own atmosphere).

Technical complexity (ages 12+)

Children may be comfortable explaining how the moon is held within the Earth's gravitational field; how it may have been formed from the Earth's own material as a result of impact; how we see only one side of the moon from Earth; that we can jump higher on the moon because its mass is one-sixth that of Earth; that the absence of atmosphere may be linked to the moon's low gravity; and may be able to calculate the gravitational force of objects using a formula. This is getting closer to 'the full story'.

Abstract manipulation (ages 18+)
Students may begin to manipulate complex, abstract concepts to address issues such as the moon's potential as a base for space exploration, or moon-based mining, or its part in the geological history of our solar system. They may have a sense of not ever 'getting the full story' as they recognise that we are always incorporating new information and solving new problems.

When you compared beliefs and explanations, what examples did you find of an increasing sophistication in concepts about the moon? Use Snapshot 3.1 above to help your analysis. Ages are indicative only, as there is no set age at which children grasp or use a particular concept, because conceptual development occurs as a response to challenges to children's existing conceptions, interests, abilities, opportunities to learn and their educational context.

What are the practical implications of these ideas about how concepts develop in sophistication and complexity?

An important consideration is 'Should I tell children who say "You can only see the moon at night time" that they are wrong?' How might this support their development? What negative effects could result?

Most teachers will agree that you should help children recognise when their ideas do not match current views, but just saying 'You are wrong' is not good teaching practice. It is more productive to help children revise their ideas. In this case, you could get children interested in trying to work out how it's possible to see the moon during the day. Productive follow-up activities could include helping children look out for the moon during the day, keeping an observation diary and predicting when the moon will be visible—you can use the moon setting and rising times available online or in the newspaper—or making and using a model to show how the moon goes around the Earth all the time and the relative position of the sun.

Review your notes and select which of the suggested follow-up activities would be most appropriate for each of the children (or adults) you interviewed to support each to develop a more sophisticated understanding about the moon's orbit. Your chosen activities need to consider the children's existing conceptions, as it is not good

Snapshot 3.2: Conceptual development of the use of a match

Response	Analysis
An 8–9-year-old child focused on some of the steps in the events: 'You stroked it on the brown and then it got a burn'.	• 'stroked' is the best word known to describe that particular action (some children will use 'wipe') • 'brown' is a way of indicating the location • 'got a burn' suggests this child has experience with flames and burning (for example, young children may say 'burnies' because it's the warning their parents use) but the child does not use the terms 'flame' or 'hot'.
An 11–12-year-old child explained, in simple terms, the series of events and used some scientific terminology: 'You striked it hard there and the friction made a flame'.	• 'striked it hard' indicates awareness that a certain amount of effort is required for the match to light, probably because a child at this age has used matches (for example, on camping trips) • 'there' is a way of indicating the location • 'friction made a flame' suggests the child has an emerging understanding of the connection of friction and flames, but does not explain why there is a connection • the child does not use the terms 'heat', 'hot' or 'energy'.
A 15–16-year-old child used terms and concepts from science classes and incorporated aspects that are not visible to the eye. He focused on why rather than a description of step-by-step events: 'Friction	• economy of expression—this child sounds like a science textbook, with precise phrasing and complex causal connections such as 'between . . . and . . . will make . . . and cause' • use of specialised terminology—'striking plate' • 'the head' suggests the child recognises that the match head (not just the match) is the specific part linked to the flame

between the match head and the striking plate will make the head combust and cause a flame'.

- 'combust' is a scientific term related to 'burn', 'ignite' and 'heat' and suggests the child has some understanding of 'friction' and combustion
- the child does not mention involvement of oxygen, chemical change, or activation temperature which is dependent on the heat which resulted from friction.

practice to provide children with highly sophisticated explanations which are not appropriate to their development level.

EXPLANATIONS DEVELOP AS CHILDREN MATURE

Key idea: As children's conceptual knowledge becomes more sophisticated, their explanations of events become more complex.

Children learn about burning and flammable materials (as part of the Energy and Change outcome) because heat energy is an important part of our daily lives, as we heat or cool our homes, cook food and burn fuel in our cars. Incorporating science concepts into explanations of how these things happen is a component of many science programs. In this activity, asking children to explain why a match burns will provide you with evidence of how explanations become more complex as children mature.

Activity 3.2: Using a match

Try this activity by yourself first, then with children of different ages, for example, 8–9-, 11–12- and 15–16-year-olds.

Take a match out of a matchbox and light the match. Let it burn for a few seconds, and then blow it out. Then ask, 'Why do you think the match does that?'

To help your comparison of children's different explanations, use the analysis of responses in Snapshot 3.2 as a guide (Hickey, 1999). Look for broad patterns of similarities and differences in how children explain the event. In each case, children may include grammatical approximations, such as *striked*, or technical terms such as *combust*. Remember not to provide a specific term or ideas (for example, don't ask 'Do you think it will burn?') but encourage the child with general prompts using only their actual words (for example, 'You said *burn*—can you tell me more about *burn*?').

What are the practical implications for teaching and planning learning activities? A critical aspect is that you recognise the differences between the three age-related examples provided, and then use this understanding to plan how to support children to make their own explanations more complex.

One practical suggestion is for you to use scientific terminology in your discussions with children. Use words such as *flame*, *burn*, *combust*, *ignite*, *energy* and *smoke* and encourage children to gain confidence in using the terms. Another strategy is to provide models of explanations, by recording children's explanations on a whiteboard or chart for others to share. A third strategy is to ask children to compare some explanations (written or oral) and discuss which are more plausible, detailed and accurate and use specific, appropriate scientific terminology.

REVISING THE RULES

Key idea: Children develop their own 'rules' about the world, based on their experiences. A new experience is added comfortably to their rules or, if it does not fit, then the rules are revised.

Sometimes, a new experience is counterintuitive, because it does not match what you expect. For example, children typically expect oil to sink when it's poured into water but are forced to rewrite their own rules when they see it sitting (floating) on top. These rules may relate to the 'properties' of materials (part of the Natural and Processed Materials outcome). A material's properties are what it is made of (for example, plastic, metal, ceramic), what it is like (for example, hard, soft, elastic, brittle, dense, transparent, reflective or flammable) and what it can or cannot do (for example, conduct electricity, insulate for temperature control, regain its shape after deformation).

In this activity, balloons are used as the stimulus to elicit children's concepts about gravity and gases, falling and floating. The activity deliberately includes an event which may be unexpected, to stimulate children's revision of their existing rules about how the world works. Children's responses are linked to the concepts they hold about the properties of materials. With the balloon stimulus, children often involve properties of gases (heavy or light) or weight (of the string or balloon) in their responses. See if you can watch a child's face as he or she experiences, for the first time, what happens when a balloon does not do what is expected!

Activity 3.3: What will happen?

Try this activity by yourself first, then with children of different ages, for example, 4-, 8- and 11-year-olds. Have one balloon blown up with air and tied with a long string. Have a second helium-filled balloon, the same size, and with the same length of string, out of sight in another room.

Give the child the first balloon. Ask 'What can you tell me about your balloon?' Then, 'What do you think (predict) will happen if you let go of your balloon?' and (after the event) 'Why do you think that happened?' Then repeat the interview with a helium-filled balloon (use the string to retrieve it from the ceiling).

To help your comparison of children's responses, use the analysis below as a guide. Try to identify the concepts children are using, what rules they are applying to this situation and if they seem confident about their rules.

Concept 1: Things fall when I let go

In our example, children's responses are based on early ideas about the properties of air-filled balloons: they get bigger as you blow them

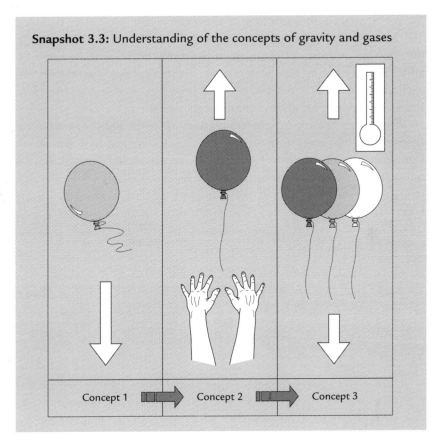

Snapshot 3.3: Understanding of the concepts of gravity and gases

Concept 1 ▶ Concept 2 ▶ Concept 3

up, they feel light and can be bounced from person to person, they float about for a while but eventually fall to the ground. Children's own experiences of accidents with glass, icecreams and toys support the development of a rule, expressed as 'Things fall down when I let go'. As their experiences multiply, children add more and more information to their concepts about what falls, what doesn't and what happens to things that fall. 'Soft toys fall, but don't break.' 'Glass falls and breaks with a lot of noise.' 'Frisbees fall if you don't catch them.'

Concept 2: Most things fall when I let go, but some things go up

Then, an event occurs which does not fit the child's rules, so the rules are updated. For example, when a child is used to air-filled balloons, then experiences a helium-filled balloon, it's time to rewrite the rules! When let go, a helium-filled balloon shoots upwards surprisingly quickly and will either hit the ceiling and stay there or, if outside, disappear out of sight. This new evidence (acquired through experience and observation) requires the child to develop a new rule: 'Most things fall when I let go, but some things go up'.

Children's reasoning about differences in how balloons behave indicates their science concepts. For example, children may talk about how 'gravity makes one balloon fall but it's not strong enough for the other balloon' or make their own generalisations—'things fall if they are heavy, but float in the air if they are light' or 'it's gravity but I don't know why'. Some may focus on less relevant aspects such as colour ('silver balloons are better than red ones') or prior experiences ('we had one of those balloons').

Concept 3: Whether it floats depends on what gas is in the balloon, air temperature and the weight of the string

As new concepts and terms are learned, children develop their understandings of the properties of materials and, in turn, revise their rules or expectations.

Children are no longer surprised by the sudden ascent of the helium-filled balloon (they have seen it before) but will often try repeated explanations, until they develop one they feel makes the most sense. They refine their ideas and try out explanations based on the 'amount of air makes it light or heavy': for example, 'The helium floats up because it's got not much air in it and the normal balloon falls because it's got a lot of air in it'. This is a useful explanation, but is not the full story. Children may attempt to use density to explain the event: 'Helium is less dense than air, so a balloon filled with helium will float in air'. Children may include understandings about pressure and temperature as they continually modify their ideas based on increasingly complex knowledge. For example, 'In a cold room, helium balloons will sink, because the gas becomes more dense; but if the room gets warmer, the density of the gas inside the balloon reduces, and the balloon will go up'.

What are the practical implications of this key idea? Teaching primary science is more about providing an environment where children feel confident to try out new ideas and attempt their own explanations, rather than waiting for the 'right answer' from the teacher. So it is important for teachers initially to accept inaccurate ideas and give children sufficient experiences and time to refine their ideas. After all, when children learn to walk, we don't label the crawling period as 'wrong', but as a necessary developmental stage. Similarly, teachers may need to suppress their immediate reaction to 'explain' and instead develop strategies that will support children to develop their own ideas at their own pace, by providing challenging, realistic, practical, issues-based investigations with plenty of opportunity for discussion and hands-on experimentation.

As adults, we continue to develop, refine and revise our ideas, as we compare, accept and incorporate new evidence into our view of things. Most of us reach a stage where our personal, conceptual theories of how the world works are stable, reliable and useful, and only occasionally are our ideas challenged, or proved inaccurate. Consider how you would feel if you believed that you inherited a gene for your big nose from your grandma and then discovered by watching television that there is no such thing as a single gene for a big nose? Wouldn't you say, 'Isn't that interesting' or 'I never knew that'? Your experience of seeing models of genes coding for the production of complex proteins and hearing that many genes and proteins as well as the environment are involved in the development of a nose would challenge you to develop a more complex view of genetics.

INTERLINKING CONCEPTS TO GET THE FULL STORY

Key idea: Children's understanding of scientific phenomena can develop from a focus on effects, to one that involves an increasing number of scientific concepts, to an eventual understanding that involves many interlinked concepts.

Children learn about the seasons during primary school, often as part of the Life and Living outcome. In the junior years, the focus is on people, plants and animals and the effects of changing seasons on activities and natural cycles (for example, many animals respond to winter by hibernation). Later, the focus may be on an explanation of why the seasons occur, involving the movement of the Earth around the sun. For older children, the focus may be more technical, with a mathematical component, and include an analysis of global weather patterns and their effect on global economy.

Activity 3.4: Why do we have the seasons?

Try this activity by yourself first, then with children of different ages, for example, 5-, 8-, 11- and 16-year-olds. Give each child a piece of A4 paper and a selection of pens and textas, and ask, 'Can you do a drawing that shows why it is that we have seasons?' Then ask, 'Can you tell me about your drawing and why it is that we have the seasons?'

Use Snapshot 3.4 to guide your analysis of children's responses. Focus on tracking:

- which concepts are used by younger children (for example, cold and hot, sun and rain) and if these focus on effects on themselves or nature;
- which additional concepts are introduced by older children (for example, revolution around the sun, proximity to the sun) and if there is a greater focus on explanation of the event; and
- repeated, attempted explanations that the child recognises are complex but still incomplete (for example, concepts of angles, area, and heat and light energies).

Teachers use state and national curriculum documents to guide their planning, including outcome statements which describe conceptual development (see Chapter 4). These documents track the increasing complexity of understanding within process and conceptual domains during the compulsory years of schooling.

Try these tasks to develop a practical familiarity with conceptual development in curriculum documents. Remember that outcomes are written in broad terms, and are usually not specific to a particular topic.

Task 1   Compare the three examples in Snapshot 3.4 of children's concepts about the seasons with your own state science outcomes.

Snapshot 3.4: Children's conceptions of the seasons

Child 1

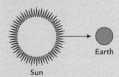

'We have seasons so we get winter when it's cold and summer when it's hot.'

This child (8 years old) had a simple understanding of differences in the seasons. He related them to personal comfort or activities, such as 'going swimming in the summer'. There was a causal connection included with 'the rain comes from the clouds' but no explanation of what caused clouds, or how they carry rain.

Child 2

Earth

Sun

'We have done this at school and I know it's to do with the Earth going around the sun and when it's closer to the sun we get the summer and when it's further away we get the winter.'

This child (12 years old) retrieved information from instruction. Some aspects conform to current scientific explanation: for example, that the Earth is 'going around the sun'. However, other aspects were not consistent with currently held explanations: it is not the Earth's closeness to the sun that causes the seasons, but differences in the angle of the sun's rays. This is a misconception frequently held by both children and adults.

Child 3

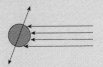

'Well it's complicated but it's to do with the sun and the tilt of the Earth. If the rays are straight on then it's summer but if they are at an angle, then it's winter. It's the light from the sun that's important. It's about the size of the land the light falls on that makes it that season. In this one [diagram] it's the northern hemisphere and winter and in the southern [hemisphere] it would be summer.'

This child (16 years old) attempted a complex answer that involved abstract concepts of light and heat, and rays that are 'straight on' or 'at an angle'. He set up multiple causal connections between sun, tilt, angles, light, land area and the hemispheres, but did not explain them succinctly. There was evidence of incomplete understanding (the diagram shows northern summer and southern winter) and the role of heat as well as light is not made clear.

Can you estimate which level of description best fits the responses for Child 1, Child 2 and Child 3? What features of the children's responses were critical in your estimates (for example, number of concepts used, the sophistication of explanations, the accuracy of ideas, or the use of technical terms)?

Task 2   Look at the children's responses from your own interviews. Try to assign levels to their explanations. What features of the responses did you use to describe how children's concepts develop? Did you discover additional features (for example, a child's apparent confidence, lack of certainty, language complexity, or differences to your own understanding)?

Task 3   When would it be possible for a primary school child to give an explanation similar to Child 3? If this child was in your class, would you be surprised, worried or enthusiastic? Would you say he didn't need any more help, get him to do independent research, or ask him to share his understanding with the class?

Task 4   Examine the box on p. 59. What level of description fits it best? Is it close to the 'full story' (a sophisticated explanation)? What more would be needed?

Some teachers feel challenged by the range of science topics and sophistication of concepts they are asked to teach in primary school (Hickey & Schibeci, 2000), but there is a range of strategies designed to meet this challenge. First, this activity illustrates how you can write

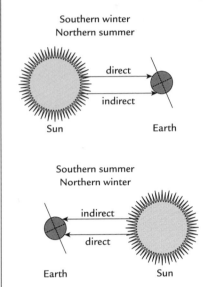

The child or adult may specify that the sun provides both light and heat; that a major determinant of seasons is the amount of sun's heat on the Earth, which varies due to the Earth's axis. During winter, the sun appears lower in the sky, daylight hours are shorter, heat rays are at a low angle, so heat is spread over a larger area. The effect is lower temperatures. In summer, when the sun is overhead the heat rays fall directly onto a smaller area of Earth's surface, which results in higher temperatures and longer days. Heat influences ocean currents as water moves due to convection currents, and affects pressure gradients through winds that move heat energy around the world. The child or adult may include the equinox or solstice and comment on the 'apparent movement' of the sun, or novel experiences such as the 'land of the midnight sun'.

or draw what you already know about a topic, then find out the concepts and understandings that some children in your class already have, then compare your findings to curriculum documents to gauge the range of levels of understanding you can expect from your class. You can develop your own understanding by researching science books and collaborating with other teachers who are familiar with the topic. Bulk loans from your school and local libraries will identify what is typically included in children's books for the topic at a range of conceptual levels. Professional development courses are provided by science teacher associations, at which you can try out suggested learning activities and gauge their potential to support conceptual development for your class. You can develop a shared, problem-solving environment in your classroom so that instead of an apologetic 'I don't know', you can respond with, 'That's an interesting question. Let's find out together'. Some teachers (Schibeci & Hickey, 2004) find that their best topics are developed from their own interests (for example, you may belong to a landcare group and know a lot about endangered species and habitat conservation) or their hobbies (for example, home renovation, which involves sophisticated concepts about the properties of materials).

The goal of your preparation is not to become an expert and amaze the children with your knowledge. Your role as a teacher is to become aware of children's existing concepts, understand how these develop, and engage children in quality learning activities. It's important to see yourself as a learner as well, so you are confident to support children to further develop their own science concepts.

SUMMARY

In this chapter, you have developed four key ideas about science teaching and learning. You have an understanding of how children's concepts of natural phenomena such as the moon are characterised by increasing sophistication or complexity as they build on their experiences. You have listened to their explanations of an event, for example, a burning match, and identified the key features to consider as you analyse their responses, and plan how to support further development. You are aware of how children use rules, or have expectations of how the world works, and how these rules are revised when they meet the unexpected, such as a balloon that flies up! You have developed an appreciation of how understanding complex phenomena such as the seasons requires children to involve multiple science concepts. Finally, you have considered how to embed these ideas in practical teaching strategies, with the goal not of getting children to 'learn the facts' but to continue to develop their own concepts about the world around them, as they 'get the full story'.

REFERENCES

Australian Academy of Science. (1994). *Primary investigations*. Canberra: Australian Academy of Science.

—— (2005). *Primary connections*. Canberra: Australian Academy of Science.

Hand, B., & Prain, V. (Eds.). (1995). *Teaching and learning in science: The constructivist classroom*. Marrickville, Vic.: Harcourt Brace & Co.

Hickey, R. (1999). *The influence of teachers' content knowledge and pedagogical content knowledge in science when judging students' science work*. Doctoral thesis, Curtin University of Technology. Retrieved September 28, 2006, from http://adt.curtin.edu.au/theses/available/adt-WCU20020522.155413/.

Hickey, R., & Schibeci, R.A. (2000). Primary teachers' conceptions of 'chemical'. *Australian Science Teachers' Journal, 16*(2), 33–38.

Kruszelnicki, K. (2002). *Dr Karl's collection of great Australian facts and firsts*. Sydney, NSW: Angus & Robertson.

Ross, K., Lakin, L., & Callaghan, P. (2000). *Teaching secondary science: Constructing meaning and developing understanding*. London: David Fulton Publishers.

Schibeci, R.A., & Hickey, R.L. (2004). Dimensions of autonomy: Primary teachers' decisions about involvement in science professional development. *Science Education, 88*(1), 119–145.

CHAPTER 4
A GUIDE TO SCIENCE CURRICULUM DOCUMENTS

Vaille Dawson
Edith Cowan University, Western Australia
Grady Venville
University of Western Australia, Western Australia

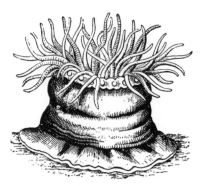

An Anemone.

OUTCOMES

By the end of the chapter you will:
- understand the key features related to teaching, learning and assessment of science curriculum documents in each state and territory;
- understand the range of scientific concepts typically taught in Australian primary schools; and
- be able to identify topics that will enhance student understanding of scientific processes and concepts.

INTRODUCTION

Commencing your first job as a beginning teacher can be a daunting prospect. After the excitement of actually getting a job settles down, there are a myriad of tasks to fulfil. Armed with your university degree and your passion for both teaching and children, you must plan a program of teaching and learning that will cater for the needs of your students. A question uppermost in your mind is likely to be: What do I actually have to teach these children? The prospect of teaching science may be of particular concern because it seems difficult (even to you) and requires equipment that may be unfamiliar.

In some schools, especially larger ones, there may be well-documented teaching programs that set out a sequence of teaching and learning activities. Other schools may have nothing. However, it can be difficult to pick up another teacher's program and use it without understanding why particular outcomes, learning activities or assessments are specified. It is a bit like following a knitting pattern without having a picture of the whole garment. Who knows what the final product will look like? You need to be aware of what your students might already know and what they will need to know by the time they leave you. The most important source of information about *what* to teach and *how* and *why* will be the government-mandated curriculum documents for your state or territory. Read on.

Ultimately, what you teach in your science lessons will be influenced by a number of factors including:

- guidelines from state and territory curriculum documents (the focus of this chapter);
- the priority given to science compared to other curriculum areas in your school and state/territory;
- the resources, time and budget allocated to science in your school (the focus of Chapter 8);
- teacher expertise and interest in science;
- teacher familiarity with learning activities in science (addressed in Chapter 6);
- teacher beliefs about the importance of teaching science;
- student factors such as ability and previous science experiences; and
- community and parental expectations.

We begin with a recent historical perspective on the development of the current science curriculum documents for each state and territory and then a summary of their key features. Website addresses are provided to allow you to access further information on support documents for your state or territory (see Table 4.1). While this chapter does not aim to provide a checklist of topics or a list of compulsory science content that should be taught in primary school it does provide an overview of the concepts and topics that are taught. The topics and concepts are organised into early childhood (kindergarten to Year 3) and middle childhood (Years 4–6/7).

WHAT IS THE PURPOSE OF A SCIENCE CURRICULUM DOCUMENT?

The science curriculum documents in all Australian states and territories provide an overview of the discipline of science with the overall aim that school students will achieve the knowledge, understanding, skills, values and attitudes of science so that they can participate fully in society. The documents are intended to be used by teachers in the planning of learning activities, teaching programs and assessment of their students' learning in science. The curriculum frameworks and other support documents also provide information on the scope (breadth and depth) and sequence (order) of scientific concepts, process skills such as communicating and investigating, and other science areas such as science in society and careers.

AN HISTORICAL PERSPECTIVE

In 1989, the Australian Education Council, which comprised the state, territory and federal education ministers, commissioned the development of national statements and profiles for eight broad curriculum areas, one of which was science. This led to the publication of *A Statement on Science for Australian Schools* (Curriculum Corporation, 1994a) and *Science—A Curriculum Profile for Australian Schools* (Curriculum Corporation, 1994b). (For economy of words we will refer to these two documents as the *National Science Statement* and *Profile*.) The development of the *National Science Statement* and

Profile involved Australia-wide consultation with practising science educators at the primary, secondary and tertiary levels, parents, industry groups, professional associations, community groups and other stakeholders. The *National Science Statement* and *Profile* provide a framework for curriculum development: the science learning area was divided into four concept-based strands—Life and Living, Natural and Processed Materials, Energy and Change, and Earth and Beyond. A fifth process strand was called Working Scientifically.

The findings of contemporary educational research on teaching, learning and assessment in science were considered. Thus, both curriculum documents were underpinned by a social constructivist and student-centred approach to teaching and learning. They moved away from a rigid syllabus style to a more flexible approach that would cater for students of varying abilities and aspirations. It was recognised that learning is an ongoing and active process, that students construct their own knowledge and that prior knowledge and experiences will influence student learning. (See Chapter 2 for a discussion on how children learn.) It was recognised that students progressed at different rates and, for this reason, rather than specify concepts according to age or year group, eight levels of achievement that reflected student learning were described. The term 'learning outcomes' was proposed so that the emphasis moved away from what the teacher taught to what the students actually knew, understood, could do and valued.

When the state and territory education ministers met with their federal counterpart, to consider the *National Science Statement* and *Profile*, each state and territory decided (for largely political reasons) to develop their own curriculum documents. In the last decade, these documents have evolved divergently to produce the current dynamic documents which will continue to change as new community and political demands are made. For example, in 2005 New South Wales introduced a new syllabus in Years 7 and 9. In Western Australia, eight new science courses of study will be introduced to Years 11 and 12 in 2008 (Curriculum Council, Western Australia, 2005c). In Queensland, a new *Queensland Curriculum, Assessment and Reporting Framework* (Queensland Studies Authority, 2007) will be implemented in all schools from 2008.

A COMPARISON OF AUSTRALIAN STATE AND TERRITORY SCIENCE CURRICULUM DOCUMENTS

The purpose of this part of the chapter is to provide an overview and comparisons of the key features of the science curriculum documents for each of the six states and two territories. We will attempt to highlight what we recognise as similarities and differences between the documents by comparing them with each other and the *National Science Statement* and *Profile*. For each state and territory, the name of the education authority, the web address, the curriculum document title and its scope are summarised in Table 4.1.

Science in the curriculum

All state and territory curriculum documents except for Tasmania's are structured around discipline-based learning areas similar to the *National Science Statement* and *Profile*. Western Australia (WA), Northern Territory (NT) and Queensland (Qld) have eight 'learning areas': The Arts, English, Health and Physical Education, Languages (Languages other than English: LOTE), Mathematics, Science, Society and Environment, and Technology (and Enterprise). Victoria (Vic.) has a similar discipline-based approach, but the learning areas are called 'key learning areas' or 'KLAs'. The Australian Capital Territory (ACT) has nine 'areas of learning', the extra one being English as a Second Language (in addition to English and LOTE). In New South Wales (NSW), the K–6 curriculum has seven KLAs—English, Mathematics, Human Society and Environment, Science and Technology, Personal Development, Health and Physical Education, and Creative Arts. From Years 7 to 10 in NSW, there is a syllabus of 28 courses including some mandatory English, Mathematics and Science (see Table 4.2). The Tasmanian curriculum lists 18 'essential' learnings that are the desired outcomes of education. These outcomes will be taught within the curriculum areas of English/Literacy, Mathematics/Numeracy, Science, Information and Communication Technology (ICT), Society and History, Health and Wellbeing, Vocational and Applied Learning, and Arts.

The structure of science within the curriculum documents

Table 4.2 summarises the key organisational features of the curriculum documents from each state and territory. When we look at the way the discipline of science is structured and represented it is evident that the state and territory curriculum documents have evolved considerably from the *National Science Statement* and *Profile*. The *National Science Statement* and *Profile* are structured into four concept strands of Earth and Beyond, Energy and Change, Life and Living, and Natural and Processed Materials. There is also a process strand, Working Scientifically. Most states and territories incorporate something comparable with these four concept strands, although varying terminology is used. For example, ACT has the 'Energy and Change conceptual strand'; in WA and Qld it is the 'Energy and Change outcome'; in Vic. the 'Physical Science strand'; in SA the 'Energy Systems strand'; and in NT the 'Energy and Change element'. In the NSW K–6 syllabus the conceptual ideas represented in the strands seem to be considerably different when compared with those of the other states and territories because of the inclusion of technology with science. The 'Physical Phenomena strand' and 'Built Environments strand' contain most of the content in relation to Energy and Change (see Table 4.2). In the NSW Year 7–10 syllabus, energy content is present in the Knowledge and Understanding outcomes.

Processes of science

The Working Scientifically process strand from the *National Science Statement* and *Profile* is represented in various forms in the individual state and territory documents. Table 4.2 shows, however, that the state and territory documents show considerable diversity in their representation of the process aspects of science. In WA and NT the Working Scientifically strand is divided into five outcomes (WA), or elements (NT); in Vic., Working Scientifically skills are included under the heading of Skills, Focus and Procedure; and in SA the science conceptual strands are underpinned by key ideas relating to process skills, including Investigation. In the ACT the Working Scientifically strand (or outcome) has three components, while in Qld science processes are included in the Science in Society outcome. In NSW (K–6), where Science is considered together with Technology,

Table 4.1: A summary of education authorities, names of curriculum documents, websites and document scope

Jurisdiction	Name of document
Australia	*A Statement on Science for Australian schools* (1994) *Science—A curriculum profile for Australian schools* (1994)
Australian Capital Territory	*ACT Curriculum Frameworks* (2000)
New South Wales	*Science and Technology K–6 Syllabus and Support Document* (1993) *Science Years 7–10 Syllabus* (2003)
Northern Territory	*Northern Territory Curriculum Framework (NTCF)* (2002)
Queensland	*Queensland Curriculum, Assessment and Reporting Framework (QCAR)* (2007)
South Australia	*South Australian Curriculum Standards and Accountability Framework (SACSA)* (2001)
Tasmania	*Essential Learnings Framework 1* (2002) and *Framework 2* (2003)
Victoria	*Curriculum and Standards Framework II* (2002)
Western Australia	*Curriculum Framework* (1998) *Progress Maps* (2005) *Curriculum Guide* (2005)

there are three Process Learning Outcomes including Investigating. In the Year 7–10 syllabus, there are five skills outcomes that include Planning Investigations and Conducting Investigations. In Tasmania, one of the essential learnings is Inquiry which includes components of Working Scientifically.

Education authority	Website URL	Scope
Australian Education Council Curriculum Corporation	—	K–Year 10
Australian Capital Territory Department of Education and Community Services	www.decs.act.gov.au	P–Year 10
New South Wales Board of Studies	www.boardofstudies. nsw.edu.au	K–6 and Years 7–10
Department of Employment, Education and Training (Northern Territory)	www.deet.nt.gov.au/ education/ntcf	Transition– Year 10
Queensland Studies Authority	www.qsa.qld.edu.au	Years 1–10
Department of Education, Training and Employment, South Australia	www.sacsa.sa.edu.au	R–Year 10
Department of Education, Tasmania	www.ltag.education.tas. gov.au	0–16 years
Victorian Curriculum and Assessment Authority	www.vcaa.vic.edu.au	Prep–Year 10
Curriculum Council of Western Australia	www.curriculum.wa. edu.au	K–Year 12

Progression of learning

All state and territory curriculum documents recognise the developmental nature of science knowledge and skills and that learning outcomes can be achieved at a range of increasingly complex and sophisticated levels (Table 4.2). The *National Science Statement* and

Table 4.2: A summary of key features of each jurisdiction's science curriculum documents

Jurisdiction	Organisation of science learning area	Levels/bands/ stages of achievement
Australia	Four concept strands • Earth and Beyond • Energy and Change • Life and Living • Natural and Processed Materials One process strand • Working Scientifically	Eight levels (1–8) of achievement
Australian Capital Territory	One process strand (or outcome) and four conceptual strands (or outcomes) each with three components Working Scientifically • Investigating • Using Science • Acting Responsibly Earth and Beyond • Earth, Sky and People • The Changing Earth • Our Place in Space Energy and Change • Energy and Us • Transferring Energy • Energy Sources and Receivers Life and Living • Living Together • Structure and Function • Biodiversity and Change Natural and Processed Materials • Materials and Their Uses • Structures and Properties • Reaction and Change	Seven levels (1–7) for each of the five outcomes
New South Wales	*K–6 Science and Technology* Six Knowledge and Understanding learning outcomes • Built Environments • Information and Communication	Six stages Stage 1 (K–2) Stage 2 (2–4) Stage 3 (4–6) Stages 4–5 (7–10)

- Living Things
- Physical Phenomena
- Products and Services
- The Earth and its Surroundings

Three Skills learning outcomes

- Investigating
- Designing and Making
- Using Technology

Three Values and Attitudes learning outcomes

- Towards Themselves
- Towards Others
- Towards Science and Technology

Years 7–10 Syllabus

Seven Knowledge and Understanding
objectives (not listed)

Five Prescribed Focus Areas

- History of Science
- Nature and Practice of Science
- Applications and Uses of Science
- Implications of Science for Society and the
  Environment
- Current issues, Research and Development
  (five outcomes)

Models, theories and laws, structures and
systems, interactions (seven outcomes:
not listed)

Five Skills objectives

- Planning Investigations
- Conducting Investigations
- Communicating Information and
  Understanding
- Scientific Thinking
- Working Individually and in Teams
  (ten outcomes)

Values and attitudes (five outcomes:
not listed)

Stage 6 (11–12)

Northern Territory	Two science strands	Key growth point 1

Working Scientifically (five elements)

- Planning
- Investigating
- Evaluating
- Acting Responsibly

Key growth point 1
Key growth point 2
Key growth point 3
Band 1
Band 2
Band 3

Jurisdiction	Organisation of science learning area	Levels/bands/ stages of achievement
	• Science in Society Concepts and Contexts (four elements, each with three outcomes) • Earth and Beyond • Energy and Change • Life and Living • Natural and Processed Materials	Band 4 Band 5 Beyond Band 5
Queensland	Five outcomes • Science and Society • Earth and Beyond • Energy and Change • Life and Living • Natural and Processed Materials	Six levels (1–6) plus foundation and Beyond Level 6
South Australia	Four Science strands with key ideas (process) and two outcomes for each strand • Earth and Space • Energy Systems • Life Systems • Matter	Standard 1 (Year 2) Standard 2 (Year 4) Standard 3 (Year 6) Standard 4 (Year 8) Standard 5 (Year 10)
Tasmania	Five essential learnings and culminating outcomes • Thinking—Inquiring and Reflective Thinkers • Communicating—Effective Communication • Personal Futures—Self-directed and Ethical People • Social Responsibility—Responsible Citizens • World Futures—World Contributors 18 key element outcomes e.g. • Inquiry • Investigating the Natural and Constructed World • Understanding Systems	Five standards for each key element outcome Standard 1 (end of K) Standard 2 (end of Year 2) Standard 3 (end of Year 5) Standard 4 (end of Year 8) Standard 5 (end of Year 10)
Victoria	Four science strands, each with two substrands Biological Science • Living Together, Past, Present and Future	Six levels of achievement over 11 years Levels 1 and 2, one

	• Structure and Function Chemical Science • Substances: Structure, Properties and Uses • Chemical Reactions Earth and Space Sciences • The Changing Earth • Our Place in Space Physical Science • Energy and its Uses • Forces and their Effects	science strand, Levels 3–6, four science strands
Western Australia	Four conceptual learning outcomes and five process outcomes Understanding Concepts • Earth and Beyond • Energy and Change • Life and Living • Natural and Processed Materials Working Scientifically • Investigating • Communicating Scientifically • Science in Daily Life • Acting Responsibly • Science in Society	Eight levels of achievement plus foundation for each of the four conceptual outcomes and investigating over 12 years

Profile described eight levels of achievement in science from Level 1, the most basic or simple understandings and skills, to Level 8, the most complex and sophisticated understandings and skills. Some states and territories have retained this basic structure. For example, WA's curriculum document describes eight levels of achievement, with the addition of a foundation level at the basic end of the continuum. ACT describes seven levels for each of the five science outcomes; Victoria describes six levels of achievement (for one science strand from Levels 1 and 2 and for four strands from Levels 3 to 6); and Queensland six levels, as well as a foundation level and 'Beyond Level 6'. NSW describes six stages, NT has three growth points as well as five bands and 'Beyond band 5', and SA and Tasmania each describe five standards for various year groups as documented in Table 4.2.

AN OVERVIEW OF AUSTRALIAN STATE AND TERRITORY SCIENCE CURRICULUM DOCUMENTS

Australian Capital Territory

The *ACT Curriculum Frameworks* (ACT Department of Education and Community Services, 2000) describes the curriculum for preschool to Year 10. It begins with a definition of science and a rationale for its importance as a school subject. The curriculum's underlying beliefs and assumptions, including a constructivist perspective, personal relevance, science for all, an investigative and holistic approach, and the importance of values, attitudes and scientific language are made explicit. In addition, the importance of gender equity, environmental education, information communication technology (ICT), language, multiculturalism, special needs education, work education, an Australian perspective and an Australian indigenous perspective are emphasised. As listed in Table 4.2, the learning area is organised into five strands, one process strand of Working Scientifically and four conceptual strands—Earth and Beyond, Energy and Change, Life and Living, and Natural and Processed Materials. Each strand has three components.

Each strand is elaborated for five bands of schooling—early years (P–Year 1); lower primary (Years 1–4); upper primary (Years 4–7); high school (Years 7–10); and post-compulsory (Years 11–12). In addition, examples of student achievement for each of the five outcomes are described for seven levels.

New South Wales

In New South Wales, there is a separate syllabus and support document for K–6 Science and Technology (Board of Studies, NSW, 1993) and a new science syllabus for Years 7–10 (Board of Studies, NSW, 2003) which was implemented in 2005/06. The *Science and Technology K–6 Syllabus and Support Document* (Board of Studies, NSW, 1993) contains learning outcomes, content, links to other learning areas and a support document containing units of work, suggested teaching strategies and resources. Under the heading of Knowledge and Understanding are six content strands—Built Environments, Information and Communication, Living Things, Physical Phenomena, Products and Services, and The Earth and its Surroundings.

There are three skills outcomes—Investigating, Designing and Making, and Using Technology—and three Values and Attitudes: Towards Themselves, Towards Others and Towards Science and Technology. Each outcome is elaborated at Stage 1 (K–Year 2), Stage 2 (Years 2–4) and Stage 3 (Years 4–6). In addition, there is information about catering for diversity with information on teaching girls and boys, Aboriginal students, students from diverse cultural/language backgrounds, students with disabilities and/or learning difficulties, and talented students.

Northern Territory

The *Northern Territory Curriculum Framework* (Department of Employment, Education and Training, Northern Territory, 2002) describes student learning from Transition to Year 10 and a major theme is that the science learning area is intended to develop scientific literacy through the integration of two strands, Working Scientifically (which has five elements of Planning, Investigating, Evaluating, Acting Responsibly and Science in Society) and Concepts and Contexts (which has four elements of Earth and Beyond, Energy and Change, Life and Living, and Natural and Processed Materials).

Each of the two strands and the associated nine elements are described as outcomes for Key Growth Point 1, 2 and 3, for Bands 1–5 and Beyond Band 5. Indicators or specific examples of appropriate activities for each element are listed for each growth point and each band. Within the indicators, links to other learning areas and to indigenous languages and culture are identified.

Queensland

The Queensland science syllabus (Queensland School Curriculum Council, 1998) covers Years 1–10 and begins with a rationale of the importance of science and a description of how science contributes to lifelong learning. The role of science in the cross-curricular priorities of literacy, numeracy, life skills and futures is outlined.

There are five strands—Science and Society, Earth and Beyond, Energy and Change, Life and Living, and Natural and Processed Materials. Each of the five strands contains core (essential) learning outcomes and discretionary outcomes to provide depth and breadth.

These are divided into a foundation level and seven levels with the seventh titled Beyond Level 6. For each of the outcomes, core content for Years 1-10 is specified. Working Scientifically is not assigned levels but is described separately under the aspects of Investigating, Understanding and Communicating with a list of components.

South Australia

The *South Australian Curriculum Standards and Accountability Framework* (Department of Education, Training and Employment, SA, 2001) has an accompanying *Draft R–10 Science Teaching Resource* (Department of Education and Children's Services, SA, 2003). There are four curriculum bands—early years (R–Year 2), primary years (Years 3-5), middle years (Years 6-8) and senior years (Years 8-10). Within the science learning area are four conceptual strands—Earth and Space, Energy Systems, Life Systems and Matter. Each strand has two outcomes. Working Scientifically process skills are integrated throughout the conceptual strands. Across the four bands, key ideas for each strand are related to process skills (for example, Investigating and the Influence of Science on Society). There are also examples of content to illustrate each of five standards. When used for curriculum planning, teachers must also take account of essential learnings—literacy, numeracy and the use of ICT.

Tasmania

The new curriculum in Tasmania, *Essential Learnings Framework 1 and 2* (Department of Education, Tasmania, 2002; 2003a), is currently in the implementation phase and will be fully implemented in 2008. Essential learnings is a distinct move away from disciplines, although it is expected that teachers will still teach in traditional subject areas. The *Essential Learnings Framework 1* (Department of Education, Tasmania, 2002) describes learning from birth to 16 years and lists five essential learnings—Thinking, Communicating, Personal Futures, Social Responsibility and World Futures; and five associated culminating outcomes—Inquiring and Reflective Thinkers, Effective Communication, Self-directed and Ethical People, Responsible Citizens and World Contributors. From these are derived 18 key element outcomes. A further document, *Essential Learning Outcomes and Standards*

(Department of Education, Tasmania, 2003b), describes five standards for each of the 18 outcomes to be achieved from K to 10. There are also illustrative examples of performance which describe in broad terms examples of learning experiences which may assist the teacher with planning specific learning experiences and monitoring student progress. Where is science in the curriculum outcomes and standards? A search reveals some familiar aspects of science within the key element outcomes of Inquiry, Reflective Thinking, Investigating the Natural and Constructed World, and Creating Sustainable Futures.

Victoria

The Victorian *Curriculum and Standards Framework II* (Victorian Curriculum and Assessment Authority, 2002a) includes the *Science Key Learning Area Overview* (Victorian Curriculum and Assessment Authority, 2002b). This document describes four conceptual strands of Biological Science (with substrands of Living Together, Past, Present and Future, and Structure and Function); Chemical Science (with substrands of Substances: Structure, Properties and Uses, and Chemical Reactions); Earth and Space Science (with substrands of The Changing Earth and Our Place in Space); and Physical Science (with substrands of Energy and Its Uses and Forces and Their Effects). There are six levels (1–6) and Level 6 extension. At Levels 1 and 2, the outcomes are combined into one science strand, while from Levels 3 to 6 there are two substrands. At Levels 3 and 4 there is one learning outcome for each substrand and two outcomes for Levels 5 and 6. For each strand at each level, there is a curriculum focus which contains a list of skills, processes and procedures that describe Working Scientifically skills.

Western Australia

The Western Australian *Curriculum Framework* (Curriculum Council, WA, 1998) covers kindergarten to Year 12. The science learning area statement begins with a rationale and definition of science and then describes nine learning outcomes: four conceptual outcomes—Life and Living, Natural and Processed Materials, Energy and Change, and Earth and Beyond; and five process outcomes—Investigating, Science in Society, Science in Daily Life, Acting Responsibly and Communicating

Scientifically. Important curriculum principles are outlined and include an encompassing view of curriculum, explicit values, inclusivity, a flexible, developmental and collaborative approach and integration. Cross curricular links and opportunities to integrate the other seven learning areas are identified.

In addition, there are two support documents that are designed to be used by teachers in their planning and monitoring of student learning: *Progress Maps* (Curriculum Council, WA, 2005a) and the *Curriculum Guide* (Curriculum Council, WA, 2005b). The *Progress Maps* describe the outcome of Investigating and the four conceptual outcomes at eight levels of increasing complexity as well as at foundation level. Examples of the types of knowledge and understanding displayed by students at each level are described. The *Curriculum Guide* provides specific examples of content and skills for four stages of development (early childhood, middle childhood, early adolescence and late adolescence) for all nine outcomes.

KEY SCIENCE CONCEPTS AND TOPICS

It is beyond the scope of this book to list every science concept and topic that should be taught in primary school science. However, our analysis of the curriculum documents reveals common areas within most of the science documents. They are summarised in Table 4.3. The list is not exclusive and is intended to provide an overview for the pre-service or beginning teacher.

Table 4.3: Key scientific concepts typically taught in primary school science

Concept	Concept area	K–Year 3	Years 4–7
Earth and Space Science	Earth forces and materials	Weather patterns and influences on human lifestyle Effects of wind, water on environment and people	Structure of Earth (including atmosphere) Oceans, tides and currents Mining processes Characteristics of minerals Effects of catastrophic events (e.g. cyclones, droughts, bushfires)

			Water cycle Erosion and weathering
	Relationship between Earth, Solar System and Universe	Seasons (observable differences) Day and night sky (e.g. shadows, clouds) Visible parts of Universe (e.g. sun, moon, stars)	Effect of Earth's rotation (day and night) and tilt (seasons) Phases of moon, tides Solar and lunar eclipses Star constellations, planets, comets Space exploration (past, present and future)
	Environmental responsibility	Careful use of resources (e.g. water, sun, electricity)	Human impact on environment (e.g. salinity) Recycling of resources (e.g. paper) Local environment
Physics	Energy sources, patterns and uses	Sun is main source of energy Use of energy in everyday life (e.g. petrol, food) Use of appliances requiring energy (e.g. hairdryer, torch) Energy safety and conservation Uses of energy (e.g. light, heat, movement)	Renewable and non-renewable energy sources Energy sources and uses Characteristics and effects of forms of energy (e.g. sound, movement, heat, solar) Safety, conservation and monitoring of energy use Examples of potential (e.g. food) and kinetic energy (e.g. water waves)
	Energy transfer and change	Energy transfer (e.g. electricity to heat, light and movement) Effect of forces on shape and motion of objects	Methods of energy transfer (e.g. direct contact, levers, waves) Factors affecting energy transfer and change (e.g. number of batteries in a circuit, wire thickness)
Chemistry	Properties of materials	Objects are made of different materials which are suited to their use	Materials can be used to make a range of objects depending on use

Concept	Concept area	K–Year 3	Years 4–7
		(e.g. steel for cars and bridges, rubber for tyres)	Solids, liquids and gases have certain properties
		Materials can be grouped according to their properties (e.g. solid, liquid and gas)	Materials may be grouped according to observable characteristics (e.g. solubility, hardness)
		Materials may be natural (e.g. wood) or processed (e.g. paper)	Materials may be changed chemically to produce a new product (e.g. bauxite to aluminium, corrosion)
		Handle materials safely (e.g. kitchen and laundry products)	Handle and dispose of materials safely (e.g. flammable material)
	Interactions and changes of materials	Materials can change from one form to another (e.g. ice to water) by heating, cooling, folding, dissolving	Solids, liquids and gases will change state with addition or removal of heat although amount of heat will depend on material
		Some changes can be reversed (e.g. dissolving salt in water) while others can't (e.g. paper to ash)	Differences between physical and chemical change
			Components of mixtures influence use
			Separation of mixtures (e.g. filtration, evaporation)
			Factors affecting reaction rates
			Chemicals can adversely affect humans (e.g. lead) and the environment (e.g. greenhouse gases)
Biology	Inter-dependence of living things	Living and non-living things comprise an environment	Living things need living and non-living factors to survive
		Living things depend on each other for survival	Food chains and food webs show transfer of energy and feeding relationships
		Living things have needs from the environment	Biodiversity
			Respiration and

	for survival (e.g. food, water, shelter, air) Living things live in different types of habitats Care of animals and plants	photosynthesis are life-sustaining processes Humans can change the environment in a positive or negative way (e.g. by introducing plants and animals or by protecting endangered species) Characteristics of Australian plants and animals
Structure and function	Living things have features that are suited to their habitat Living things differ from non-living things Living things can be grouped according to similar characteristics (e.g. birds have wings) Living things have structures that help them survive (e.g. plants have leaves to make food)	Living things are adapted to their habitat for survival (e.g. fish have fins and gills) Characteristics of living things (e.g. they move, respond to stimuli, produce wastes and are made up of cells) Living things are classified according to external and internal structural features (e.g. plants/animals, vertebrates/non-vertebrates, flowering/non-flowering) Structure and behaviour of living things aid in survival (e.g. camouflage, hibernation)
Reproduction and change	Living things grow and change over their life span (e.g. kitten to cat, egg to insect) Living things reproduce and produce similar offspring Living things have changed over time (e.g. horses)	Living things that reproduce asexually (through one parent) (e.g. using cuttings, runners or spores) produce identical offspring Living things that reproduce sexually (two parents) produce similar but not identical offspring (e.g. humans, flowers) Life cycle of plants, insects (e.g. beans and silk worms)

Concept	Concept area	K–Year 3	Years 4–7
			Factors influencing life cycle (e.g. nutrients, water quality, soil) Living things are influenced by and adapt to changes in their environment over long periods of time

SUMMARY

In this chapter we have provided a summary of the key features of the curriculum documents that guide the teaching, learning and assessment of science throughout Australia. It is important to note that while these curriculum documents do provide guidance for teachers, the documents themselves cannot guarantee good teaching. Rather, they are a guide for the teachers and it is the teacher who interprets them. What and how we teach will inevitably be influenced by our own beliefs and values about science, teaching experience, understanding of science, science pedagogy, pedagogical content knowledge, resources available, student age, school culture, ability, interest and aspirations of students, and parental and community expectations.

ACKNOWLEDGMENT

This chapter is a revised version of a paper published in *Teaching Science* Vol. 52.2 and is reproduced with kind permission of the publishers, Australian Science Teachers Association.

REFERENCES

ACT Department of Education and Community Services. (2000). *ACT curriculum frameworks*. Retrieved August 29, 2006, from http://www.decs.act.gov.au/publicat/acpframeworks.htm.

Board of Studies, New South Wales. (1993). *Science and technology K–6 syllabus and support document.* North Sydney, NSW: Board of Studies, New South Wales. Retrieved August 29, 2006, from http://www.boardofstudies.nsw.edu.au.

—— (2003). *Science Years 7–10 syllabus.* North Sydney, NSW: Board of Studies, New South Wales. Retrieved August 29, 2006, from http://www.boardofstudies. nsw.edu.au.

Curriculum Corporation. (1994a). *A statement on science for Australian schools.* Carlton, Vic.: Curriculum Corporation.

—— (1994b). *Science—a curriculum profile for Australian schools.* Carlton, Vic.: Curriculum Corporation.

Curriculum Council, Western Australia. (1998). *Curriculum framework for kindergarten to Year 12 education in Western Australia.* Osborne Park, WA: Curriculum Council, Western Australia. Retrieved August 29, 2006, from http://www.curriculum. wa.edu.au.

—— (2005a). *Curriculum framework progress maps, science.* Osborne Park, WA: Curriculum Council, Western Australia. Retrieved August 29, 2006, from http://www. curriculum.wa.edu.au.

—— (2005b). *Curriculum framework curriculum guide, science.* Osborne Park, WA: Curriculum Council, Western Australia. Retrieved August 29, 2006, from http://www.curriculum.wa.edu.au.

—— (2005c). *New Western Australian Certificate of Education* (WACE). Retrieved August 29, 2006, from http://newwace.curriculum.wa.edu.au/pages/home.asp.

Department of Education, Tasmania. (2002). *Essential learnings framework 1.* Retrieved August 29, 2006, from http://www.ltag.education.tas.gov.au/references.htm.

—— (2003a). *Essential learnings framework 2.* Retrieved August 29, 2006, from http://www.ltag.education.tas.gov.au/references.htm.

—— (2003b). *Essential learnings outcomes and standards.* Retrieved August 29, 2006, from http://www.ltag.education.tas.gov.au.

—— (2003c). *Introduction to the outcomes and standards.* Hobart, Tas.: Department of Education, Tasmania. Retrieved August 29, 2006, from http://www.ltag.education. tas.gov.au.

Department of Education and Children's Services (DECS), South Australia. (2003). *Draft R–10 Science Teaching Resource.* Hindmarsh, SA: DECS Publishing. Retrieved August 29, 2006, from http://www.sacsa.sa.edu/companion.

Department of Education, Training and Employment (DETE), South Australia. (2001). *South Australian curriculum standards and accountability framework.* Seacombe Gardens, SA: DETE Publishing. Retrieved August 29, 2006, from http://www.sacsa.sa.edu.au.

Department of Employment, Education and Training, Northern Territory. (2002).

Northern Territory curriculum framework, learning area—science. Retrieved August 29, 2006, from http://www.deet.nt.gov.au/education/ntcf/index.shtml.

Queensland School Curriculum Council. (1998). *Years 1–10 science syllabus.* Spring Hill, Qld: Queensland School Curriculum Council. Retrieved August 29, 2006, from http://www.qsa.qld.edu.au.

Queensland Studies Authority. (2007). *Queensland curriculum, assessment and reporting framework.* Retrieved March 20, 2007, from http://www.qsa.qld.edu.au/qcar/index.html.

Victorian Curriculum and Assessment Authority. (2002a). *Curriculum and standards framework II.* East Melbourne, Vic.: Victorian Curriculum and Assessment Authority. Retrieved August 29, 2006, from http://www.vcas.vic.edu.au.

—— (2002b). *Science key learning area overview.* East Melbourne, Vic.: Victorian Curriculum and Assessment Authority. Retrieved August 29, 2006, from http://www.vcas.vic.edu.au.

PART II
IMPLEMENTING THE ART OF TEACHING PRIMARY SCIENCE

CHAPTER 5
PLANNING TO TEACH PRIMARY SCIENCE

Christine Preston
University of Sydney, New South Wales
Wilhelmina Van Rooy
Macquarie University, New South Wales

OUTCOMES

By the end of this chapter you will:
- understand the benefits of different levels of planning in primary schools;
- be able to analyse a science teaching program from a constructivist perspective; and
- be able to plan and prepare lessons that support a constructivist approach to teaching and learning science.

INTRODUCTION

Planning in primary science is important for two reasons: first, to ensure that students are involved in appropriate, stimulating learning experiences; and second, to ensure that such learning experiences are combined and sequenced in a coherent, logical manner. Effective teachers think about what science knowledge and/or skills their students need to acquire, how classroom activities can be presented to engage their students in science and how the required learning goals can best be achieved. As indicated in Chapter 4, curriculum documents generally provide an overview of the required content, skills and values for teaching and learning science. Teachers use these documents as the basis for planning topics and individual lessons that link everyday classroom experiences to state or territory goals and requirements (Hinde-McLeod & Reynolds, 2003).

When planning, teachers also need to think about the way they will teach science. The science teaching approaches most widely supported by current research are based on the theory of constructivism (see Chapter 2 for more details). The next chapter, Chapter 6, describes a variety of teaching approaches that have been found to be successful for the teaching and learning of science. While variations of constructivist teaching–learning models exist all follow a similar pattern of:

- eliciting students' pre-existing ideas;
- selecting activities that challenge students' ideas and assist them to further develop their ideas; and
- reflecting on students' ideas and learning (Skamp, 2004).

Planning for teaching science is a complex task that requires teachers to draw on an array of knowledge and resources, creativity and logic. Carefully constructed planning documents give teachers confidence and generally improve their teaching of science. This chapter outlines methods of planning to teach science in primary schools that are consistent with a constructivist framework and provides examples of successful planning documents.

TYPES OF PLANNING DOCUMENTS FOR PRIMARY SCIENCE

In primary schools, planning for science teaching occurs at different levels and involves school personnel with varying levels of responsibility.

> In a staff meeting at the end of the year, the whole staff discuss changes that might be made to the whole-school plan for science. I go away and make any modifications that are necessary and then give it back to the staff for them to approve or make more suggestions. So the whole-school plan for the new year is usually ready by the end of the previous year and the teachers can start thinking about their own planning. That also gives me the chance to clean out and update the science storeroom knowing what kinds of topics students in each year group are going to be doing. (primary school Science Coordinator)

There are various levels of planning for science teaching and learning in a primary school that result in different kinds of planning documents (Barry & King, 1998). Typically there is a whole-school plan. Each teacher also prepares an overview for their class for the year as well as a plan for each topic or theme (sometimes referred to as a program or 'forward planning document'). Teachers also prepare lesson plans (sometimes referred to as a 'session plan' or 'planned learning experience'). The following sections consider each of these documents and the processes required for developing them in more detail.

Whole-school planning

In primary schools where science teaching is well-established, you will be able to consult a whole-school plan before planning your own teaching. The whole-school plan provides an overview of the school's approach to teaching science and sequences topics for all year levels. It takes into account school-wide issues, and provides for continuity of experience and cognitive development. This ensures that what children learn in earlier years is built upon in subsequent years. A whole-school plan also eliminates needless repetition. Without careful planning and cross-checking children may be taught things they have already studied or not be provided with comprehensive

Table 5.1: Contents of a whole-school plan

Introduction	Provides vital information about the way in which science is taught in the school (e.g. particular teaching approaches or school-based goals) and the factors that will impact on your teaching and students' learning. These might be whether the school is single sex or co-ed, whether science is taught by the classroom teacher or a specialist science teacher and whether the school has special resources or facilities such as a science room.
Aims and rationale	These are based directly on state or territory curriculum documents but reworded to suit the specific school context or goals (e.g. increasing the participation of girls in science).
Scope and sequence	Shows all of the science topics to be taught in all year groups across the school year (see Snapshot 5.1).
Outcomes	Provides evidence that all of the state or territory curriculum document outcomes are planned to be met. This usually takes the form of some kind of curriculum map across all years in the school that shows in which years and in which topics outcomes are taught.
Assessment	Details of planned formal assessments and reporting requirements.

learning opportunities. Table 5.1 gives you an indication of the information a whole-school plan might contain.

Snapshot 5.1 is from a primary school in New South Wales and shows a typical layout for a whole-school scope and sequence. At a glance any teacher in the school can see how the topics she or he will be responsible for planning and teaching fit into the school science plan. For example, if you were a kindergarten teacher in this school you would commence teaching 'What's Alive' in Term 1, followed by 'People and Weather' in Term 2 and so on.

Teacher planning

Once you get down to teaching and learning activities, each teacher is responsible for their own planning. University lecturers, principals and

Snapshot 5.1: School scope and sequence, science K–6

Year	Kindergarten	Year 1	Year 2	Year 3	Year 4	Year 5	Year 6
Term 1	What's Alive?	Growing Up	Toy World	Earth's Oceans	Living Together	Light Up My Life	Local River Study
Term 2	People and Weather	People and Products	Energy Around Us	Invisible Forces and Energy	Earth and Space	Materials and Substances	Systems and Services
Term 3	Push me, Pull you	People, Time and Weather	Wet and Dry Environments	People and Substances	Machines and Uses	People and Flight	Electricity and Energy Changes
Term 4	Let's Experiment	Movement and Energy	Living Things and Weather	Human Body	Lego™ Robotics	How Plants Work	Earth Changes

teachers all use different terms to describe a teacher's plan. For example, a teacher's plan is commonly called a program or forward planning document. Marsh (2004) describes a program as a 'teacher's creative representation or interpretation of a curriculum. It should follow the broad principles of the curriculum but the emphasis and combinations of activities will represent each teacher's judgements about what they consider to be important for their particular class' (2004, p. 79). In other words, a teacher's plan is a document that you, as a teacher, create to communicate the ways that you decide to teach your class. In primary schools, it is normal practice for class teachers to submit their plans to the principal or other supervisor early in the school year or at the beginning of each school term. The teacher's plan provides official documentation of your work for a school year so it must be readable and intepretable by others including replacement teachers.

It is important to note that in primary schools, science is sometimes taught as a separate subject, but often is integrated with other learning areas such as mathematics, English or technology. Moreover, many Australian teachers now use 'rich tasks' or problem-based approaches to guide what is to be learned in each of the learning areas, including science. These factors can create some tension when teachers begin to develop their science plan. On the one hand, science should be considered as an integral aspect of the entire learning that students experience during a year. On the other hand, it is important that the science doesn't get lost or overlooked and that teachers give careful consideration to and are explicit about their students' science learning. Part of being an excellent teacher is being able to plan so that specific science outcomes are achieved and, at the same time, the students' science learning experiences contribute to and do not seem disjointed from their wider education. Regardless of whether science is taught as a separate subject or integrated with another learning area, the principles of planning for science outlined in this chapter remain highly relevant.

Planning documents vary enormously depending on the context and philosophy of the entire school community, the beliefs of the teacher who writes them and the specific purposes for which they are written. This means that there is no set framework for writing a teacher's plan. As a general rule of thumb, structure and content should be guided by usability. An effective program is one that is used rather than one that sits on the shelf gathering dust.

Commonly, a teacher's plan consists of an overview and a plan for each unit of work during the year. Units of work (or topics) are usually between five and ten weeks in duration. Primary school students should be engaged in a minimum of one hour of science per week, but we would strongly encourage teachers to commit considerably more of their weekly schedule to the study of science.

The teacher's overview

The overview to a teacher's science plan for a school year is likely to consist of a class scope and sequence, a class timetable, a needs analysis and a specific program for each unit of work.

Class scope and sequence   Just as the whole-school plan shows the sequence of topics for all years, a teacher's plan will show the order of topics (or units of work) to be taught and might include an overview in the form of a brief description of each topic (see Snapshot 5.2).

Class timetable   An outline of when and how often your class has science is useful to help you plan and to organise resources and equipment for your science lessons. While this may seem basic, it conveys daily routines and is a vital communication tool for other teachers who may take your class when you are absent.

Needs analysis   Teaching and learning activities should be based on the learning needs of the children in your class. Teachers usually conduct a needs analysis (also known as a 'situational analysis' or 'class profile') to determine the makeup of their class group. This might include information about the students' gender, ages, achievement levels, special needs, interests, medical backgrounds (especially allergies) and any other factors that may influence teaching and learning. While you need to wait until you meet your students to collect such information, you can plan beforehand how to find out about your students' needs and what information you think will be important.

Plans for units of work   Your teaching program will also include individual plans for each unit of work. These are the most detailed part of your teaching program and the next section explores planning units of work in detail.

Snapshot 5.2: Class scope and sequence for kindergarten

Term	Topic	Outline
1	What's Alive?	This unit aims to enhance young children's rudimentary understandings about and interest in living things. The characteristics of living things will be investigated through the direct observation of different plants and animals. Children will design and make animal enclosures that will provide for an animal's needs.
2	People and Weather	Daily events including the weather provide an interesting topic for capturing young children's innate curiosity about the world around them. This topic will assist children in identifying changes that occur due to natural processes and how humans cope with such changes. Children will associate words that they may know already to types of weather, different times of the day and different times of the year.
3	Push me, Pull you	Children will investigate forces in their daily lives, explore how torches work and experiment with a range of toys to identify forces involved and energy needs. Electrical energy from batteries and power points will be introduced with an emphasis on safety with electricity. This topic forms the basis of development of forces and energy that will underpin future learning in secondary school.
4	Let's Experiment	This topic is designed to develop children's investigation skills through a range of activities. Children will observe different materials using a magnifying glass, test soap bubble mixtures and conduct simple experiments involving dissolving and melting. The children will also investigate rocks by observing different types of rocks through magnifying glasses and make their own rocks with readily available materials.

PLANNING UNITS OF WORK

When planning units of work you are planning a sequence of lessons that will develop logically children's knowledge and understanding of content, skills, values and attitudes across a period of some weeks. This means you need to consider the order of activities carefully. Which ones should be at the beginning of the unit and which should be towards the end? You also need to plan for ongoing assessment throughout a unit of work so that evidence is collected to evaluate children's progress.

> I feel so satisfied when I've finished planning a unit of work for the term. It really gives me the big picture of the outcomes the students are going to achieve and how they will achieve them. It gets me excited too, because I keep looking for different and exciting things that the students can do that will get them through that learning journey. I always modify the plan as I go along. (Year 5 teacher)

Teachers view this as the most important part of their planning because it describes *what* the students are going to learn and *how*. Teaching programs are individual, often idiosyncratic, and, therefore, vary in their components. An example of a five-week program for a unit of work called 'What's Alive?', written for a Year 2 class, is provided in Snapshot 5.3.

This program includes the outcomes for the state curriculum documents as well as two goals that the teacher wants to achieve in the Life and Living conceptual outcome and the Investigating outcome. The program is set out as a table, with each lesson taking up a row and columns representing the specific learning outcomes for each lesson, the activities the children will be involved in, the techniques the teacher plans to use for the management of the classroom, the resources and materials required as well as assessments and other comments.

Analysing a unit of work from a constructivist perspective

Careful analysis of the program provided in Snapshot 5.3 shows that the teacher has developed a logical sequence of activities that is consistent with a constructivist model of teaching such as the 5E model described in Chapter 6 (Bybee, 1997). The first phase of the 5E model is Engage—students' interests are engaged in the topic and their prior knowledge is elicited. The first two lessons involve the students observing the environment outside the classroom in the school grounds and drawing pictures of things they think are 'living' and 'not living'. This is a form of diagnostic assessment in which the students' pre-existing ideas about the topic are elicited and made public in the form of posters.

During lessons three to seven, the students are involved in a number of activities including investigating plant growth, observing kittens, analysing a television program on baby animals, reflecting on their own growth and comparing humans, cats and plants in terms of growth and reproduction. These lessons intermingle the Explore, Explain and Elaborate phases of the 5E model as students are provided with hands-on experiences and are required to construct explanations and representations of this developing conceptual

Snapshot 5.3: Five-week plan for a unit of work called 'What's Alive?'

Outcomes

Life and Living Outcome Level 1: The student understands that people are examples of living things and that, like all living things, they change over time.

Life and Living Outcome Level 2: The student understands that needs, features and functions of living things are related and change over time.

Investigating (conducting) Level 1: The student carries out activities involving a small number of steps, observes and describes.

Investigating (conducting) Level 2: The student observes, classifies, describes and makes simple non-standard measurements and limited records of data; and uses independent variables that are usually discrete.

Teaching goals
1. To develop students' understanding of living things to encompass animals, plants and humans by using reproduction and growth as criteria.
2. To develop students' ability to observe, measure and record information.

Lesson	Learning outcomes	Activity	Classroom management	Resources	Assessment/comments
1–2	• Students demonstrate their understanding of the phrase 'living thing'	• Observation of outside environment and drawing of living and non-living things in the environment	• Partners to hold hands outside • Individuals draw pictures back in class	• A3 paper, coloured pencils	• Check pictures for misconceptions • Remind students (with note home) to bring baby photos for lesson 4
3	• Students understand that plants grow from seeds • Students observe, measure and record plant growth with string	• See growing activity	• Students to work in small groups to set up growth investigations • Measurements to be taken on Mon and Thurs of every week with string	• Activity sheet with list of resources in teacher's file	• Check group records of string stuck to pin-up board • Need lesson later to go through string records of plant growth with students and come to some conclusions

Lesson	Learning outcomes	Activity	Classroom management	Resources	Assessment/ comments
4	• Students understand that humans have babies and grow	• 'Show and tell' baby pictures and story about growing up	• 4 groups of 7 students. Each child has 3-minute show and tell	• Students to bring photos from home	• Check students talk about babies and growing up • Send note home about costs and include response sheet from parents if students have allergies
5–6	• Students understand that cats have babies and grow • Students understand the different ways that baby animals need care	• Guest to bring mother cat and kittens (birds or dogs) to class • Video—*Harry's Practice* special on baby animals	• Quiet observation of mother and baby cats • Whole-class discussion about cats (birds, dogs) having babies and growing up • Video analysis of what they observed about baby animal care	• Parent and baby animal required (check with parent) • Contact RSPCA for information on animal care • Video in school library	• Check during discussion that students understand that cats grow and that kittens are baby cats • Question students about care of animals they observed in video

7	• Students identify that a cat, a plant and a human (all living things) have 'babies' that grow • Students make comparisons between things	• Drawing of babies and adult humans, cats and plants	• Individual work drawing	• A3 paper, coloured pencils	• Check understanding that human babies, kittens and seeds are all 'babies'
8–9	• Students understand that all living things have babies and grow • Students classify and make comparisons between things	• Students classify pictures of things as either living or non-living based on whether they have babies and grow • Reflection on lesson 1 pictures	• Group work discussing and classifying pictures of things • Individual work renaming pictures from lesson 1 as living and non-living	• 5 sets of pictures of living and non-living things, laminated	• Groups to compare and discuss how they classified the pictures

understanding. The ideas about the concept of living are extended from humans to the less familiar contexts of plants and animals. Depending on the time available, the students' interests and achievement levels and the local resources, this phase could be greatly extended. Ideas for extension might include an excursion to a local zoo or a tree nursery, a visit to the school from a travelling animal farm or further experimentation with plant growth.

The final two lessons in the program involve the students in a form of summative assessment where they classify a set of pictures as living and non-living things and reflect on their original posters. This is the Evaluate stage of the 5E model because students are required to re-present their understanding and reflect on their learning journey. It is also an opportunity for the teacher to collect evidence about achievement of outcomes.

The planning process

When developing a sequence of activities, teachers can draw on a vast range of resources including curriculum documents and support materials, student textbooks and teacher manuals, multimedia including the internet, software and television, excursion sites, professional science teachers associations, colleagues' knowledge and old programs (see Chapter 8 for more information about resources for teaching science). As you become more experienced as a teacher, you will refine your knowledge of these resources and how best to use them when planning. Teachers may choose either to follow closely a sequence of lessons suggested by a reputed curriculum package such as *Primary Connections* (Australian Academy of Science, 2005) or *Primary Investigations* (Australian Academy of Science, 1994), to modify these programs based on their students' needs and interests, or plan their own program from scratch. Planning might revolve around a significant local event such as the potential destruction of a nearby wetland due to the development of a car park, a recent electrical storm or an excursion to an exhibition at a local museum. Planning is rarely a linear process, starting from the first lesson and finishing at the last. It is a more complex process—the teacher goes back and forth, modifying and changing the order of lessons, re-writing outcomes and adding assessments until she or he is generally satisfied.

An effective program is a working document. Practising teachers plan in advance, teach their lessons and then annotate their program to show any variations between the planned and the actual. Teachers invariably deviate from what was planned because their teaching and learning circumstances are changing constantly. Many primary schools require teachers to sign their annotated teaching programs as an official record of what has been taught to the children.

The bigger picture of a plan for a unit of work

As well as a sequence of activities, a plan for a unit of work also could include an introduction, an overview of students' prior ideas, details of assessment, a section for the teacher's self-evaluation and a list of detailed information about references.

Introduction   A contextual statement briefly outlining the content and types of learning activities your topic program is designed around. Outcomes should be listed to provide evidence of clear links to your state or territory's curriculum documents.

Prior ideas   Taking into account your students' prior knowledge and understanding—including possible alternative conceptions—is a vital aspect of planning. Teachers can find out students' prior ideas by looking back at previous science topics and by incorporating diagnostic assessment into the program.

References   A reference section allows you to acknowledge information and activity sources, reminds you about the original source of your ideas and enables you to refresh your memory about an activity or background knowledge.

Self-evaluation   Professional teachers continually seek to improve their practice, therefore self-evaluation is a vital part of your teaching program. This is where you include reflective comments, possible improvements and changes. Evaluation becomes effective when you develop a formal written record of your reflections, thoughts and ideas before developing subsequent programs.

Assessment   Details of planned diagnostic, formative and summative assessments, reporting requirements and class records also form part of your teaching program (see Chapter 9 for a more detailed discussion on assessment). Assessment strategies are best planned in advance to ensure that you focus on assessing the outcomes that you want the children to achieve. This enables you to include a variety of assessment strategies and to select those that best suit the learning activities included in your topic program.

LESSON PLANNING

As a pre-service or beginning teacher your first experience in planning will most likely be writing individual lesson plans. Lesson plans are essential for the beginning teacher as they set out the timing and sequence of activities during the lesson.

> The lesson plans I wrote just gave me so much more confidence. Even though I didn't actually read every bit of it while I was teaching, I knew what was there and I could refer to the important bits and make sure that everything I wanted to do in the lessons got done. (4th year Bachelor of Education student teacher)

An example of a lesson plan is provided in Snapshot 5.4. Analysis of this example shows that it was designed to ensure that the lesson had an effective introduction and conclusion, was based on children's prior understanding, had some hands-on practical activity (concrete experience) and allowed for discussion time to assist the children to make sense of their observations and develop a scientific understanding to explain their findings. The structure conveys clear links with curriculum document outcomes, specific learning outcomes (written as indicators), resources and timing.

There are various starting points for planning lessons, each specific to your teaching situation. During your practicum experience you might be asked to teach a lesson that is part of a particular topic, seeks to develop specific outcomes (skills and/or content), is built around an activity, or fosters the development of children's understanding of a concept. Let us assume the starting point for your lesson plan is an activity that you learned about in your teacher training and seems just perfect for your Year 3 class. Here's one way you might plan this lesson.

Snapshot 5.4: Example lesson plan

Outcome
Define a force as a push, pull or twist and give examples of the effects of forces in everyday life.

Indicators	Learning activities	Resources/time

INTRODUCTION

Identify objects that need a push or a pull or both to make them move. State that forces (push, pull, twist) can *make things move.*	Show students a tissue box, a torch and a toy car. Ask them how they can make them move (or work). Ask three volunteers to show the class. Discuss the movements each student used. Was it a push or pull? Brainstorm prior knowledge of terms—force and energy. Challenge students to suggest things in the classroom that need a push or a pull to make them move; e.g. chair, door, cupboard. Ask how a drawer can be made to move (pull to open, push to close).	Toy car, torch, tissue box. 10 mins

MAIN BODY

	Students to experiment in pairs with the provided collection of objects, as well as others they find in the room. (Students place objects in groups under the labels provided depending on whether they have to push, pull or twist them to make them work.)	Objects (toy cars, pram, wheelbarrow, torch, tissue box, peg, pull-along toy, spinning top). Labels—PUSH, PULL, TWIST. 10 mins
	Make a long-eared bat and use pushes and pulls to make its ears move.	Square paper, scissors, textas. 10 mins

Indicators	Learning activities	Resources/time
	Students pair up and use their bodies to: • push against each other; then • pull against each other; • feel the force; and • notice that they won't fall down so long as they keep applying an equal force.	Pairs of children. 5 mins
	CONCLUSION	
	Students recall the actions used (push, pull, twist), realise that this is what scientists call 'force' and that force can make things move. Ask students to draw a picture to show types of forces and what they do.	Blank worksheet page. 10 mins

First, check whether the activity aligns with the science curriculum outcomes (see Chapter 4) for this year level. If so, you can use it in your lesson. If not, you will need to find another activity that is appropriate for Year 3. List the curriculum document outcomes and write a specific learning outcome or indicator, that is, what the children will know, do or value by the end of the lesson.

Second, think about how you are going to implement this activity. How will you explain what the activity involves and what you want the children to do and think about during the activity? This is most important, as it will determine the effectiveness of the activity as a learning experience. Thorough planning—considering the specific tasks and activities the children will do, the timing and location of these activities and what you will need—will increase the likelihood of a smooth lesson. This part also involves thinking about what might go wrong or prove challenging and what you might do proactively to prevent this. Planning for classroom management (or student behaviour) at this point is much better than having to think on your feet once children start doing things you did not expect. In addition, write down questions you want to ask during discussion, list the materials and resources you will need and think about the time each part of the activity will take. You should consider organisational aspects such as where the children will be seated, how you will arrange for them to move, for example, from the floor mat to their desks and back again.

Third, how can you introduce the lesson to your class? What can you do to motivate the children, to stimulate their thinking or to get them excited about what they will be learning? How will you determine the children's prior knowledge and understanding or check the required skills for the task? Do you know something about the children's prior learning experiences that you can use as a link to this lesson?

Fourth, how will you conclude the lesson? Every lesson should have a clear conclusion. How will you sum up what has been learned in the lesson, reinforce key points and make it clear that this lesson has come to an end? Will this lesson impact on future lessons and, if so, how will you communicate this to the children?

Finally, how are you going to assess the children's learning and your teaching during this lesson? If you include a space on your lesson plan template in which you can list evidence of the children's learning you will then know to pay attention to this after the lesson. You should record actual evidence of learning here, rather than assuming that the children all learned what you expected them to. Were there any children who really stood out as being high achievers or low achievers during this lesson? Also consider your teaching: list two positive aspects of your teaching during the lesson and two points to work on to improve your teaching practice.

Using a daily planner

As you become a more experienced teacher you are less likely to put as much effort into lesson planning. Most teachers eventually use some form of 'day book' planner (see Snapshot 5.5) as a daily organisational tool, in tandem with their teaching program.

> When I first started teaching I had a file full of blank lesson plans that I filled in each night ready for the next day. Now that I've been teaching for a few years I just use a detailed program and a teacher's diary

Snapshot 5.5: Sample day book planner

3G Daily lesson planner		Monday 22 September
Period	Time	Lesson details
1	9.00	English—Procedure outline steps, group tasks (reading different types of procedural texts, sharing with class).
2	10.00	Science—Making crystals (joint construction of procedure, emphasising and comparing English and Science requirements).
	11.00	Recess

instead of lesson plans, but when I've got an important lesson or I'm teaching something new, or I'm not confident about what I'm teaching, or someone is coming into my class, I still do a lesson plan. (Year 3 teacher)

SUMMARY

Planning for the teaching of science in primary schools occurs at different levels including whole-school planning, teachers planning units of work and lesson planning. Individual teachers develop programs for units of work based on the needs of the students in their classes. Planning for effective science teaching begins with elicitation of students' prior knowledge and understanding, followed by the selection of activities that will challenge and further develop students' ideas and culminates in reflection on students' learning. Assessment of students' learning and evaluation of the effectiveness of teaching and learning activities are other vital aspects of planning for effective science lessons.

ACKNOWLEDGMENT

This chapter was based on planning materials developed by Christine Preston during her work as a specialist science teacher at Abbotsleigh Junior School. Samples have been reproduced with permission.

REFERENCES

Australian Academy of Science. (1994). *Primary investigations*. Canberra: Australian Academy of Science.

—— (2005). *Primary connections: Linking science with literacy*. Canberra: Australian Academy of Science.

Barry, K., & King, L. (1998). *Beginning teaching and beyond* (3rd ed.). Katoomba, NSW: Social Science Press.

Bybee, R. (1997). *Achieving scientific literacy: From purpose to practices*. Portsmouth: Heinemann.

Hinde-McLeod, J., & Reynolds, R. (2003). *Planning for learning*. Katoomba, NSW: Social Science Press.

Marsh, C. (2004). *Becoming a teacher: Knowledge, skills and issues* (3rd ed.). Sydney, NSW: Pearson Education Australia.

Skamp, K. (Ed.). (2004). *Teaching primary science constructively* (2nd ed.). Southbank, Vic.: Thomson.

CHAPTER 6
TEACHING STRATEGIES FOR CLASSROOM LEARNING

Denis Goodrum
University of Canberra, Australian Capital Territory

OUTCOMES

By the end of this chapter you will:
- understand the issues of scientific literacy and student learning theory and their impact on teaching strategies;
- be able to develop a student-centred approach to teaching;
- understand the different student-centred models of teaching; and
- be able to structure lessons for meaningful learning.

INTRODUCTION

The purpose of this chapter is to provide some simple and practical advice on how to improve the quality of your teaching. This will result in better learning by your students. To understand the evolving development of teaching strategies one needs to appreciate the influence of two significant forces that have emerged in recent years.

The first issue is *why* science is taught in our schools. The prevailing view used to be that school science was necessary for the initial preparation of scientists or science-related positions. Notwithstanding the importance of this consideration, today we believe the primary purpose of teaching science in our schools is to promote scientific literacy. In simple terms, scientific literacy refers to the extent to which people are able to use science in their daily life. This shift in focus has had a profound impact on both what we teach and how we teach.

The second issue to consider is our understanding of *how* people learn science. Chapter 2 explains in detail the current ideas on learning and the factors that affect learning. While there is much we still do not know about learning, the research of the past 50 years has revealed the limitations of previous teaching practices. Our present views on learning provide valuable insights into the types of teaching strategies we need to employ to provide opportunities that will result in more meaningful learning by students.

SCIENTIFIC LITERACY

There is widespread agreement within the education system and the broader community that the purpose of science education is to develop scientific literacy. The authors of a national review of Australian science teaching and learning (Goodrum, Hackling & Rennie, 2001) defined the attributes of a scientifically literate person as those shown in Figure 6.1. These attributes inform the type of learning we expect from the compulsory years of schooling. For this learning to occur, teachers must relate learning to a real-world context that is meaningful for the student. Obviously, students who seek occupations in science-related fields will pursue their interest in post-compulsory studies. All students, however, have the right to a science education that enables them to feel confident about and able

Figure 6.1: Attributes of scientifically literate people

```
                    Scientifically literate people

  are interested in and
  understand the              are able to engage        are sceptical and
  world about them            in discussions of         questioning of
                              and about science         claims made by
                              matters                   others

  can identify and                                  can make informed
  investigate                                       decisions about the
  questions and draw                                environment and their
  evidence-based                                    own health and
  conclusions                                       wellbeing
```

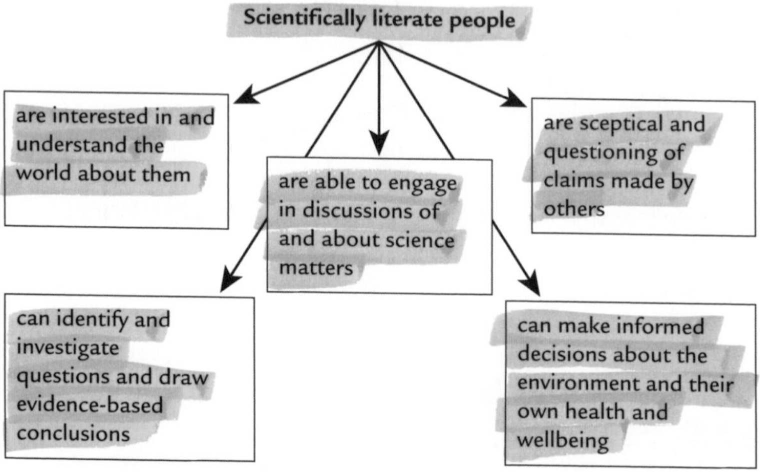

to deal with the scientific issues that impact on their lives. Primary school students have a natural curiosity about nature and the whole world of science. Our role as teachers is to promote this curiosity and our students' intrinsic interest in science.

To believe in a science education that promotes the development of scientific literacy gives rise to some expectations about the way the science is taught. Snapshot 6.1 attempts to outline some of the changes teachers should make to their practice to honour a commitment to teach for scientific literacy. In examining this snapshot, it would be wrong to suggest that one column contains only good teaching approaches and the other only poor approaches, rather it is a question of emphasis. There is a need to do more of one and less of another, if one is to teach for scientific literacy. For example, there should be less emphasis on memorising the names of scientific terms and more emphasis on learning broad concepts that can be applied to new situations, but there will still be some scientific terms that are useful for a student to know and apply.

HOW DO STUDENTS LEARN?

Our understanding of how students learn affects the way in which we teach. In previous years, teaching was dominated by the transmission

Snapshot 6.1: Teaching for scientific literacy

Teaching for scientific literacy requires:	
Less emphasis on:	**More emphasis on:**
Science being interesting for only some students	Science being interesting for all students
Covering many science topics	Studying a few fundamental concepts
Theoretical, abstract topics	Content that is meaningful to the students' experience and interest
Presenting science by talk, text and demonstration	Guiding students in active and extended student inquiry
Asking for recitation of acquired knowledge	Providing opportunities for scientific discussion among students
Individuals completing routine assignments	Groups working cooperatively to investigate problems or issues
Activities that demonstrate and verify science content	Open-ended activities that investigate relevant science questions
Memorising the name and definitions of scientific terms	Learning broader concepts that can be applied in new situations
Learning science mainly from textbooks provided to students	Learning science actively by seeking understanding from multiple sources of information, including books, internet, media reports, discussion, and hands-on investigations
Assessing what is easily measured	Assessing learning outcomes that are most valued
Assessing recall of scientific terms and facts	Assessing understanding and its application to new situations, and skills of investigation, data analysis and communication
End-of-topic multiple choice tests for grading and reporting	Ongoing assessment of work and the provision of feedback that assists learning

model. This model held that the mind of a student was empty and the role of the teacher was to fill it with scientific facts and principles. Teaching was synonymous with telling. This teacher-telling or didactic approach became entrenched in our school system with a related content-based testing regime. Even today some schools are still influenced by the simplicity of this approach, despite all we now know about how students learn.

As explained in previous chapters, today's prevalent view of learning is that students construct meaning from their previous knowledge and the new experiences or information they encounter. Learning is an active process in which learners make sense of their world by developing meaningful constructions between what they know already and the new experiences that change this knowledge. Learning, therefore, is a continual incremental process of comparing, testing and adapting. To learn new ideas and skills takes time—some learners take longer than others. A better understanding of how students learn allows us to suggest some important teaching principles that have a profound impact on the way teaching occurs.

Explanation follows experience

Unfortunately some teachers explain scientific ideas to their students without providing any contextual or experiential base for the students. The impact of such explanation is minimal. For a scientific explanation to be effective, a teacher must first create or tap into experiences upon which the explanation is based. This gives teachers a stronger chance of providing meaning to the student. There are a variety of ways teachers can either provide experiences or help students relate to previous experiences, but the most common is to provide concrete activities that allow students the opportunity to inquire and investigate. For example, for a student to develop an understanding of the concept of 'floating', the student needs to have experience with objects that float, objects that sink and sinking objects that could be made to float. Through questioning, discussion and explanation, the ideas associated with floating can be developed from the students' direct, hands-on experiences.

Recognising prior experience

As we now know, all students come to any learning situation with some preconceived ideas, which impact on how and what students learn. A teacher needs not only to appreciate what ideas students hold but how to build on this understanding.

There are a number of simple strategies you can use to identify the ideas or experiences your students have. First, at the beginning of a new topic, ask students to write a sentence or two (or draw a picture) to describe their initial understanding of the new concept. For example, in introducing a unit on energy students could be asked to write a simple sentence containing the word energy and then answer the question 'What is energy?' Another approach is to pose a problem perhaps using a picture as a stimulus. Students could discuss possible solutions, either as a class or in small groups. For example, 'Alan said he got his energy from sleeping but John said he got his energy from food. Who is right?'

As a general principle, you should always allow time to revise the previous lesson before starting a new lesson.

Student involvement

If they are to learn, students need to be interested and engaged. Good teachers continually seek to spark their students' curiosity by relating learning to current events or personal experiences. For example, a recent television show might give you the opportunity to raise an issue related to the given topic. Creative teachers make effective use of popular activities and events to illustrate teaching ideas.

Hands-on activities enhance student understanding. For example, to understand the ideas associated with electricity, students should be offered opportunities to play with batteries (dry cells) and light bulbs. The more personal and practical the involvement, the greater the potential for learning.

Student discussion

In previous years, some believed that a teacher's ability could be measured by the quietness of their class. Today we realise that quietness is not necessarily correlated with learning. This does not diminish the importance of classroom management skills but acknowledges the significance of student discussion in facilitating learning.

As students attempt to construct meaning and understanding they need to test and verify their thoughts by discussing them with their peers and their teacher. Chatting about their ideas in small groups or as a class allows students to refine and adjust the conceptual pictures they create in their minds. Ideas need to be related to evidence and views need to be justified.

This does not mean teacher explanation is not important. It is, but for maximum impact it needs to be used judiciously. An effective classroom has a balance between teacher explanation and student discussion.

The other important skill you need to develop is the ability to summarise coherently the ideas generated from student discussion. A blackboard summary of these ideas is a good way to provide worthwhile student notes. In developing this summary you can challenge and refine possible alternative conceptions or inaccurate information.

Developing conceptual understanding

Intellectual rigour is an important issue in learning. Many science educators believe rigour is not measured by the number of scientific facts memorised but rather by the depth of conceptual understanding. There is a difference between learning for memorisation and learning for understanding. If you understand a concept you can apply your understanding to a new situation. Many present-day curriculum documents outline learning outcomes in terms of developing levels of conceptual understanding rather than science topics to be covered.

It is too simplistic to classify learning into these two categories, learning for memorisation and learning for understanding, when the fact is that there are many shades of grey between the two. Learning for understanding will entail being able to remember some factual information relevant to the particular concept. As with teaching for scientific literacy, it is a question of emphasis.

To teach for conceptual understanding is more challenging than the traditional view of teaching facts and principles. Despite the challenges, most teachers try to pursue the goal of more effective learning for their students.

Questions and questioning

In traditional teaching, the driving force was teacher explanation. In inquiry-based teaching, the main engine for facilitating learning is the use of questions and discussion. To be successful, teachers need to develop an effective questioning technique. To improve your questioning technique, use the following simple but powerful skills.

Ask a balance of broad and focused questions   Broad questions such as 'What do you observe about this flower?' and 'Why are the parts of a flower arranged as they are?' stimulate student thinking. A traditional approach uses mainly narrow or closed questions designed to challenge students to recall information. There are occasions, in a discussion, when you should focus student thinking with a narrow question. Effective inquiry lessons have a balance of questions that range between broad and narrow.

Allow for sufficient wait time   Wait time is the time you, as a teacher, are willing to wait for students to answer a question. Research (Tobin, 1987) strongly indicates that a wait time of three seconds or so allows students to learn better. This time provides the opportunity for student reflection and comparison. If a student answers quickly you should allow another three seconds. Hence students will be able to think further about the question. Remember, as the teacher you control the wait time, not the students.

Use 'evaluation-free' responses   To develop a better inquiry atmosphere in a class it has been suggested that it is better to avoid comments such as 'good boy', 'great answer' and 'well done'. Rather, student responses are accepted or rejected on available evidence. An 'evaluation-free' style of responding to students results in a more normal discussion. This approach encourages independent thought and inhibits the common classroom game called 'Guessing what the teacher thinks'. (In this game, praise is bestowed on students who are successful in reading the teacher's mind rather than thinking for themselves.)

Listen   A good teacher is a good listener. By listening to a student response you are trying to understand the thinking behind the answer. As a result, you can ask more thoughtful and effective follow-up questions.

TEACHING MODELS

To translate the theories about learning into classroom practice, various authors have devised teaching/learning models that suggest ways in which teachers can organise their science lessons effectively. These models tend to emphasise aspects of the previously described principles. While there are similarities between the models, the differences reflect the varying degrees of emphasis of the principles. The following three models each have had an impact in Australia. These models normally apply to a science unit but can also be used within a lesson or a series of lessons.

The 5E model

The 5E model is a simple model that is used in the primary science curriculum resource *Primary Investigations* (Australian Academy of Science, 1994) and the more recent *Primary Connections* (Australian Academy of Science, 2005). The 5E model consists of five distinct but interconnected phases.

1 Engage Students' interests are captured through a stimulating activity or question. Students have the opportunity to express what they know about the unit topic or concept so that they can make connections between what they know and the new ideas being introduced.

2 Explore Students explore problems or phenomena through hands-on activities using their own language to discuss ideas. This exploration provides a common set of experiences upon which the new ideas can make sense.

3 Explain After the 'engage' and 'explore' experiences, explanations and scientific terms are provided to students to help them develop their ideas.

4 Elaborate Students apply what they have learned to new situations. They have discussions using the newly-acquired language to clarify their understanding.

5 Evaluate Students evaluate what they have learned and learning is assessed.

The generative learning model

This model was developed by Roger Osbourne and Peter Freyberg (1985). It proposes three distinct teaching phases.

1 Focus The teacher establishes a context by which the new concept can be explored. This context provides some motivation and interest for students. Within this phase the ideas of the students are clarified.

2 Challenge The students' ideas are challenged and compared with the evidence from the scientist's view.

3 Application The new ideas are applied to new situations and problems through student discussion and analysis.

The interactive model

The interactive model revolves around the questions of students. The following description of the approach has been developed by Faire and Cosgrove (1988). (It has been suggested that one needs to be an experienced teacher to implement this model successfully.)

1 Preparation The teacher and class select an agreed topic and seek background information.

2 Before views Students explain what they know about the topic, prompted by the teacher.

3 Exploratory activities These activities provide the basis for further questions and stimulation.

4 Student questions This is the opportunity for students formally to ask questions about the topic.

5 Investigations From students' questions investigations are identified, selected and carried out.

6 After views Students complete individual or group statements about the topic. These are compared with 'before views'.

7 Reflections During this time students reflect on what has been verified and what still needs to be determined.

Structuring an activity lesson

One of the ideas intrinsic to this chapter is that students learn more effectively from activity and experience, rather than from listening to teacher explanation. These activities may involve hands-on inquiry, discussion and information research. Some teachers believe that the move from teacher- to student-centred learning gives the teacher less of a role to play in the student activity. The opposite is true. In student-centred learning the teacher needs to help to structure situations by which students can learn more effectively through questions and relevant comments.

Every activity proceeds through three simple phases: introduction, activity and conclusion. Within any lesson this sequence may be repeated more than once.

1. Introduce the task or activity (brief)
- capture students' interest by:
  - relating to personal experiences
  - provocative questions
  - relating to previous lesson
  - provocative demonstration
  - relating to TV, movie, recent event
- outline activity explaining any necessary advanced organiser;
- explain equipment arrangements; and
- outline time and outcome expectations.

2. Student activity (main time allocation)
- assist groups or individuals with materials and their activity; and
- discuss ideas with groups and individuals, challenging them to think more deeply about what they are doing.

3. Conclude activity with discussion (allow adequate time)
- students share results of activity; and
- through questioning help students to summarise the main ideas.

The curriculum resource *Primary Connections* (2005) offers many lesson plans that illustrate this activity lesson model. These lesson plans cover a range of topics and concepts.

COOPERATIVE LEARNING AND GROUP WORK

Working in a group or team enables students to share their experiences and to consider different points of view and solutions to a problem. Cooperative learning is an approach that encourages students to work together to help them learn better. Teams develop the social skills of sharing leadership, communicating, building trust and managing conflict. These skills take time to develop but the longer-term benefits are worth the effort.

The benefits of cooperative learning include:

- *More effective learning* Students learn more effectively when they work cooperatively than when they work individually or competitively. They have a better attitude towards their schoolwork.
- *Improved self-confidence* All students tend to be more successful when working in groups and this builds their self-confidence.
- *Better class management* When students work in cooperative groups they take more responsibility for managing the equipment and their behaviour.

Students need to learn how to work cooperatively. Even though most classes offer a balance between individual, team and class activities, students need to work together regularly if they are to develop effective team learning skills.

STRUCTURING COOPERATIVE LEARNING

Use the following ideas when planning cooperative learning with your class.

- Assign students to teams rather than allowing them to choose partners.
- Vary the composition of each team. Give students opportunities to work with those who might be of different ability level, sex or cultural background.
- Keep teams unchanged for sufficient time for them to learn to work together successfully.
- If the number of students in your class cannot be divided into

teams of equal numbers, form groups of smaller rather than larger sizes. It is more difficult for students to work together effectively in larger groups.

- Consider the use of specific team jobs to help students work together.

Team jobs

If your class has limited experience of working in groups, you could consider the use of specific team jobs. Students are assigned jobs within their team and while each team member has a specific job they are all accountable for the team's performance. Each team member should be able to explain the team results and how they were obtained. It is important to rotate team jobs each time a team works together, to give students the opportunity to perform different roles. It has been suggested that colour coding (e.g. coloured wool bracelets) could be used to distinguish team jobs (Australian Academy of Science, 1994).

Possible team jobs are:

- *Manager (red)* The manager is responsible for collecting and returning the team's equipment. The manager also tells the teacher if any equipment is damaged or broken. All team members are responsible for cleaning up after an activity and getting the equipment ready to return.
- *Speaker (blue)* The speaker is responsible for asking the teacher or another team's speaker for help. If the team cannot decide how to follow a procedure, the speaker is the only person who may seek help. The speaker shares any information obtained with the team members. The teacher may speak with all team members, not just the speaker. The speaker is not the only person who reports to the class; each team member should be able to report on the team's results.
- *Director (green)* The director is responsible for making sure that the team understands the team activity and helps team members to focus on each step to be completed. When the team has finished, the director helps team members to check that they have completed all aspects of the activity successfully. The director provides guidance but is not the team leader.
- *Reports Coordinator (yellow)* The reports coordinator is responsible for ensuring that team members have completed all the

necessary reports, data collection and relevant worksheets. The reports coordinator does not necessarily report on behalf of the team. Any team member should be able to report on behalf of the team. The reports coordinator ensures each member of the team has the necessary information so that they can report to the class if required to do so.

Team skills

In addition to creating team roles that help them manage the work of small groups, it is important for teachers to help students develop the skills that make teams more cohesive and improve the learning outcomes. Teachers need to assess their students' team skills and focus on each skill that could enhance their work. The choice of skills will depend on the skill level of the particular class. It is better to focus on one skill at a time. Teachers need to identify the skill they wish to develop, and explain to the students what is expected of them and how the skill will enhance their group work and learning. In addition, teachers need to give regular feedback on students' use of the selected skills.

Skills that improve group management
- Move into your groups quickly and quietly.
- Speak softly so only your team mates can hear you.
- Stay with your group.

Skills that help groups function as a team
- Use your team mates' names.
- Look at the person speaking to you.
- Listen to others without interrupting.
- Praise others.
- Treat others politely.

Skills that enhance learning
- Contribute ideas to the discussion.
- Encourage others to participate.
- Question the ideas of others, but disagree with the idea not the person.
- Modify your ideas when provided with new information.

SPECIFIC TEACHING STRATEGIES

Concept mapping

Concept maps allow students to represent in diagram form what they know about the links and relationships between concepts. They allow students to access their prior knowledge and provide teachers with feedback on what is known or unknown and/or what is misunderstood either at a single point or over time. Concept maps are designed to increase the students' ability to organise and represent thoughts and to help with reading comprehension.

How it works When you first introduce concept maps, model their use at the whole-class level. Brainstorm ideas as a class and ask students to help group the words generated. Explain that each concept can be used only once. Link relationships with arrows or lines and talk aloud to model the cognitive processes involved. Ask students to assist in identifying and labelling the relationships between concepts.

Example When introducing the topic of energy write the word 'energy' on the blackboard with a box around it then ask, 'What other ideas are related to energy?' Write these words on the blackboard with each word in a box and a line connecting each word box. You may for example have the following words connected: energy—burning—wood—tree—plant—sun. The resulting diagram will look like an asymmetrical spider web with the word 'energy' at the centre of the web. Students who have experience with concept mapping could be asked to prepare a personal concept map of 'energy' at the start of the unit. Then ask them to repeat the exercise at the end of the unit. Students could compare their before and after ideas as outlined in the interactive model described on p. 117.

Brainstorming

The purpose of brainstorming is to generate ideas quickly—it is a creative, problem-solving strategy.

How it works Quantity is more important that quality; all ideas are accepted and not criticised. Hitchhiking (building from each others' ideas) is encouraged. Write exactly what is said—no paraphrasing.

Example   At the start of any investigation, it is a good idea to reflect on the factors or variables that could impact on the investigation. Brainstorming could be used to do this. Write the ideas on the blackboard as students call them out. Remember you should not assess the worth of any suggestion until you have written up all the suggestions.

Envoy

This strategy encourages students to learn from each other and take responsibility for their learning. It helps students to develop listening and oral skills and promotes skills in synthesising and summarising.

How it works   Students are formed into groups and then given a topic to discuss. One student from each group is selected to be the envoy. When a group has completed its discussion of the topic the envoy reports to another group and outlines what was discussed. The envoy also listens to a report from the group that he/she is visiting. The envoy then returns to his/her original group and they exchange new ideas. Each group should now have input from three groups.

Example   Divide the class into groups of, say, four and ask the students to develop an explanation for why there are no longer dinosaurs. After 15 minutes of discussion ask each group to select their envoy. Use a circular approach for sending an envoy to the next group. In this way no group will miss out.

Gallery walk

Individual student or group work is placed around the room and students are given the opportunity to view other students' work and to 'show off' their own.

How it works   Students' work is placed around the room and students are given time to circulate and view the display. Students can use the opportunity to read information prepared by others or consider the way in which the information is presented. They can collect information from the work of others or peer assess using a set of guidelines, sometimes called a rubric, that teachers prepare in advance.

Example Divide students into groups of three and ask them to design a device for using the sun to heat 200 mL of water. The students could draw their devices on large pieces of paper. The drawings are then placed on tables and the class circulates and examines the designs.

Jigsaw

This strategy provides a structure for group work and also allows students to cover a broad amount of information in a shorter period of time.

How it works Students are formed into 'home' groups of about five or six. The topic is divided up into sections and each student in the home group is given a different aspect of the topic to research. The home groups split up and the students form into 'expert' groups, in which all members of a group are researching the same aspect of the topic. Students research their aspect of the topic in the expert groups and prepare to report to their home group. They then return to their home group and take turns to report as the 'expert' on their aspect of the topic.

Example The class is divided into home groups of five students that are asked to investigate types of energy. Each member of the group is assigned a different type of energy to examine. Within the classroom are designated energy centres on particular energy types such as 'chemical energy' or 'solar energy'. After each student has received his or her energy type they go to the appropriate energy centre and find out as much as possible about that type of energy by reading the available material and/or by chatting with the other students at that centre. The students then return to their home group and each member makes a presentation to their group on their chosen topic. They could also prepare a short handout for the other members of the group.

Predict-Observe-Explain (POE)

This strategy encourages students to think more carefully and critically about scientific phenomena by challenging them to examine events that surprise them.

How it works Students are given a situation and asked to predict what happens when something is done to change that situation. As the change is implemented the students make careful observations about what happens. By comparing their observations with their predictions they are encouraged to develop explanations of what is happening.

Example The teacher sets up a demonstration in front of the class as in Figure 6.2. A ruler is balanced on the forefingers of a person's hands. Students are asked to predict what will happen when one hand (hand B) is moved towards the other hand, which stays still (hand A). After students have discussed the various possibilities ask them to write down their personal prediction. Most students will suggest the ruler will fall on the left or right side of the hand that does not move (hand A). In front of the class carry out the activity and ask students to observe closely. Discuss their observations. The ruler will move then stop then move again but will not fall. Ask the students to explain what is happening. Why does it sometimes move? Why does it sometimes not move? The explanation will introduce the ideas of friction and slipping and the factors that affect these ideas.

Figure 6.2: A ruler being balanced on the fingers of two hands

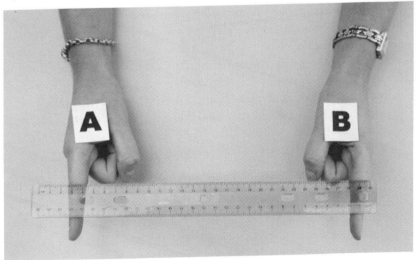

SUMMARY

This chapter begins with two questions that have a profound impact on teaching science. Why do we teach science? How do we learn science effectively? These questions lead to the ideas of scientific literacy and constructivist learning. From these ideas a series of teaching principles are implied and described. These principles are embedded in various models that are examined. Using the principles, a suggested approach to planning lessons is outlined. This chapter also provides advice on how to develop cooperative learning using a team approach and concludes with a variety of specific strategies with relevant examples.

ACKNOWLEDGMENT

This chapter is adapted from *The Art of Teaching Science* (2004), Chapter 4.

REFERENCES

Australian Academy of Science. (1994). *Primary investigations.* Canberra: Australian Academy of Science.

—— (2005). *Primary connections.* Canberra: Australian Academy of Science.

Faire, J., and Cosgrove, M. (1988). *Teaching primary science.* Hamilton, NZ: Waikato Education Centre.

Goodrum, D., Hackling, M., & Rennie, L. (2001). *The status and quality of teaching and learning of science in Australian schools: A research report.* Canberra: Department of Education, Training and Youth Affairs.

Osbourne, R., & Freyberg, P. (1985). *Learning in science.* Portsmouth, NH: Heinemann Educational Books.

Tobin, K. (1987). The role of wait time in higher cognitive level learning. *Review of Educational Research, 57*(Spring), 69–95.

CHAPTER 7
INQUIRY AND INVESTIGATION IN PRIMARY SCIENCE

Mark W. Hackling
Edith Cowan University, Western Australia

The Cicada—showing the Two Compound Eyes
and Three Simple Eyes.

OUTCOMES

By the end of this chapter you will:
- understand how an inquiry-oriented approach to primary science actively engages students in learning, stimulates their curiosity and helps them learn the skills and understandings of science;
- be able to describe types of investigations and the opportunities for learning they provide;
- be able to select appropriate scaffolds to support students' work through the phases of an investigation; and
- be able to use the scientific conventions of representing observations in the form of diagrams, tables and graphs.

INTRODUCTION

Scientific knowledge helps us understand the world around us and predict future events. Science processes and understandings help us to make sense of media reports about scientific matters, to investigate and solve problems, and to make decisions about our health and the environment. Most occupations and recreational activities require the application of science knowledge. Scientific knowledge evolves gradually as new evidence and ideas develop through inquiry and investigation. Scientific knowledge gained through carefully designed experiments with appropriate samples, adequate control of variables and appropriate tests provides trustworthy evidence that can be used to develop important public policies.

If students are to understand the nature of science and how trustworthy evidence is developed through science experiments, they must learn by inquiry and by conducting investigations. These are engaging approaches to the teaching and learning of science because students become active participants in the learning process and inquiry stimulates their curiosity and the excitement of discovery. Such approaches are also effective in developing the skills and understandings of science.

This chapter will address a number of questions related to practical work in science.

- What is an inquiry-oriented approach to learning?
- What opportunities for learning are provided by different types of practical work?
- What processes are required to complete an investigation?
- How can student learning be scaffolded and facilitated?
- How can we support students to learn the literacy skills required to represent their observations using the conventions of science?

INQUIRY-BASED LEARNING IN SCIENCE

What is inquiry-based learning? Students learn by inquiry when learning is focused on 'finding out'. Learning science by inquiry involves students asking questions then exploring and investigating natural phenomena through manipulating materials. Inquiry thus involves gaining experiences and making observations before developing explanations for those experiences. With their teacher's support, students are then able to answer their own questions.

Inquiry-based learning actively engages students in learning, it stimulates their curiosity and the excitement of discovery and it provides an authentic experience of the nature of real science, as described in Chapter 1. Moreover, inquiry-based learning facilitates the development of inquiry skills and provides personal and concrete experience of natural phenomena as a basis for developing explanations and conceptual understandings. Contemporary learning theory supports the idea that learning is most effective when students are active players in the learning process, when the learning proceeds from experiences to explanations and when the teacher helps students use their existing knowledge to develop new explanations. All of these conditions for effective learning in science are satisfied when the teacher uses an inquiry approach.

Figure 7.1 illustrates a model of inquiry learning developed by the Australian Academy of Science's (2005) *Primary Connections* Program. In this model, inquiry commences with students' own questions which provide a focus for investigating through hands-on experiences with the phenomenon of interest. The teacher uses discussion of students' observations and ideas to develop scientific explanations and representations of the observations and explanations. The teacher monitors students' discussions and representations and provides feedback to learners to help them make links to additional concepts and to raise new questions that may lead to further inquiry.

INQUIRY AND SCIENCE LEARNING OUTCOMES

An analysis of science curriculum documents throughout Australia reveals that students are expected to develop a range of learning outcomes relating to three facets of science understanding: Science as a Body of Knowledge; Science as a Way to Know; and Science as a Human Endeavour.

Science as a Body of Knowledge involves students in developing an understanding of science concepts and using these to explain and predict events. For example, students might learn the concept of light

Figure 7.1: Elements of the *Primary Connections* inquiry approach (Australian Academy of Science, 2005, p. vii; see acknowledgment at end of chapter)

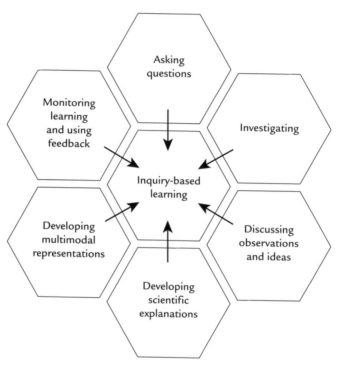

travelling in straight lines and use it to predict where shadows will be created when the sun rises over the horizon at dawn.

Science as a Way to Know involves students in using inquiry skills to pose questions, investigate, and collect and interpret observations in order to draw evidence-based conclusions. For example, students may be curious about why so few plants grow on the beach and this might lead to questions about whether beach sand is able to support plant growth. Students might investigate their question by planting seeds in pots of beach sand and pots of garden soil then observing and recording the plants' growth.

Science as a Human Endeavour involves students understanding that all people practise aspects of scientific thinking as a part of their everyday lives which contributes to their personal, economic and social wellbeing. For example, students may visit a local garden centre and see how scientific processes and understandings are used by people who propagate and grow plants.

Through developing increasingly sophisticated understandings of these three facets of science throughout their schooling, students become scientifically literate. Practical investigations provide students with opportunities to develop learning outcomes that contribute to scientific literacy, including the skills and understandings needed to conduct scientific investigations and to evaluate critically the claims made by others based on scientific evidence. Science investigations provide a rich opportunity for students to understand science as a way to know their world.

TYPES OF INVESTIGATIONS

There are three main types of hands-on investigations conducted by early childhood and primary students. These are explorations, guided investigations and student-planned investigations.

Explorations are relatively unstructured activities where students have the opportunity to 'play' with materials and gain concrete experiences of a phenomenon. They are often used to engage the learner, stimulate curiosity, raise questions and to allow students to get a feel for phenomena such as sinking and floating.

Guided investigations are strongly scaffolded by the teacher who guides the students through a sequence of steps which may involve manipulating materials, making observations or measurements, or recording, discussing and interpreting observations. Such investigations provide structured experiences of the phenomenon and lead to the collection of observations that can be used to develop explanations for the phenomenon. For example students may follow directions to test a supplied set of materials to see whether they float or sink and record the results in a table designed by the class with the teacher's guidance.

These guided investigations provide students with opportunities to practise skills of following directions and making observations; develop skills of representing observations using drawings, tables and graphs; and look for patterns in observations and interpret results. As the teacher plans the guided investigation, there is no opportunity for students to develop skills associated with formulating a research question, identifying and manipulating variables or planning fair tests.

Student-planned investigations provide students with an opportunity to plan and conduct an experiment within the context and boundaries set by the teacher. As students are required to make decisions about the experimental design and data collection, recording and interpretation, there are opportunities for student ownership, engagement and intellectual challenge. Students are also able to practise inquiry skills. Student-planned investigations are more inquiry-oriented than guided investigations and thus provide a more authentic experience of the nature of science. For example, students curious about whether beach sand can support plant growth might formulate a question for investigation and plan how they would test if garden soil is better able to support plant growth than beach sand. In their planning of their investigation students would need to consider how they would ensure that their investigation was a fair test and how they would observe or measure plant growth and record their observations.

Teachers should use a range of types of practical work so that students have opportunities to learn a wide range of understandings and skills. Variety also helps cater for individual differences in students' learning styles.

Students need to build a foundation of skills, understandings and confidence before attempting more complex investigations and projects. Using the three types of hands-on investigation outlined here, Figure 7.2 illustrates an instructional sequence that prepares students for these more demanding tasks.

STUDENT-PLANNED INVESTIGATIONS

Student-planned investigations require students to plan the experiment, conduct the experiment and collect data, process the data, communicate their findings and, finally, reflect on their investigation, evaluating the quality of their findings and the investigation methods used. The evaluation phase focuses students on what they have learned from the investigation and this can be carried forward to improved planning of the next investigation. Figure 7.3 illustrates a model of the investigation processes. In practice, these processes may not take place in the strict order of planning—conducting—processing—evaluating,

Figure 7.2: A learning sequence that prepares students for more complex forms of science investigation

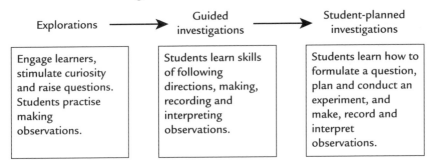

as part-way through their investigation students may realise that further planning is required to improve the measurement technique or design of the experiment.

In addition to these four phases of investigation, there are opportunities for students to communicate their findings using posters, formal written reports and oral reports. This allows them to practise communicating using scientific genres of reporting and multimodal representations combining written text, drawings, tables and graphs.

PROGRESSION IN LEARNING INVESTIGATION SKILLS

Explorations, guided investigations and student-planned investigations require progressively more sophisticated and wider-ranging skills. A developmental approach needs to be taken to planning students' learning of these skills and processes. The National Scientific Literacy Progress Map describes progression in these student learning outcomes (MCEETYA, 2005). Figure 7.4 summarises developmental levels of investigation outcomes and is based on the National Progress Map.

Many students need considerable support if they are to engage successfully in student-planned investigations and develop the investigation skills described in Figure 7.4. Teachers can support students by breaking the task into a series of steps and by modelling some of the more difficult skills.

Figure 7.3: A model of science investigation processes (Hackling & Fairbrother, 1996, p. 30; reproduced from *Australian Science Teachers' Journal* Vol. 42.4 with kind permission of the publishers, the Australian Science Teachers Association)

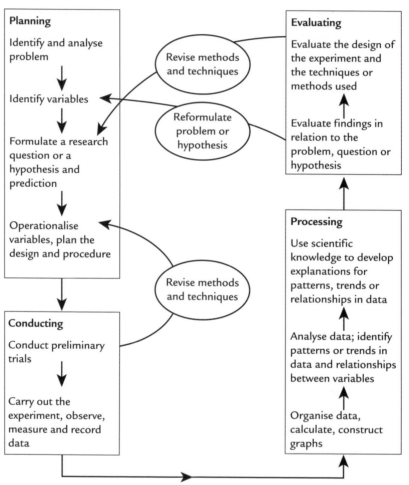

INVESTIGATION PLANNING AND REPORT SHEETS

Planning and report sheets guide students through a sequence of decision-making steps that provide structure and scaffolding to support students while leaving the students responsible for the decision-making. These scaffolds are structured by a sequence of

Figure 7.4: Progression in investigation skills over four levels (This figure is based on Table A1.1 in the *National Year 6 science assessment report: 2003*, published by MCEETYA, 2005)

Level	Planning	Conducting and processing
	Formulating or identifying investigable questions and hypotheses, planning investigations and collecting evidence.	Interpreting evidence and drawing conclusions, critiquing the trustworthiness of evidence and claims made by others, and communicating findings.
1	Responds to the teacher's questions, observes and describes.	Describes what happened.
2	Given a question in a familiar context, identifies a variable to be considered, observes and describes or makes non-standard measurements and limited records of data.	Makes comparisons between objects or events observed.
3	Formulates scientific questions for testing and makes predictions. Demonstrates awareness of the need for fair testing. Makes simple standard measurements. Records data as tables, diagrams or descriptions.	Displays data as tables or bar graphs, identifies and summarises patterns in science data. Applies the rule by extrapolating or predicting.
4	Identifies the variable to be changed, the variable to be measured and several variables to be controlled. Uses repeated trials or replicates.	Calculates averages from repeat trials or replicates, plots line graphs where appropriate. Conclusions summarise and explain the patterns in the data. Able to make general suggestions for improving an investigation (e.g. make more measurements).

Note: The national proficiency standard for scientific literacy of Year 6 students is set within Level 3.

questions and prompts. A planning scaffold suitable for Year 1–2 students is illustrated in Figure 7.5.

This is a type of scaffold that teachers can use to involve students in developing a class plan for a simple investigation. In these early years of schooling, students would then follow the class plan to

Figure 7.5: An investigation planning scaffold suitable for Year 1–2 students

---

Investigation planner

Question: Which things roll the furthest?

We will change: The things we roll.

We will observe: How far they roll.

We will keep the same: The push and the surface.

---

observe and describe or act out what happened to the objects. Students in Years 3–4 would record their observations in their science journal.

Older students in Years 5–7 would work collaboratively in their groups planning their investigation using a more complex scaffold. Figure 7.6 summarises the prompts and questions that could be included in a scaffold for upper primary students.

Scaffolds such as the one illustrated in Figure 7.6 lead students through the investigation process, eliciting from them information about their thinking and what they are doing at each stage of the investigation. Scaffolds therefore support students and the written record of the investigation provides teachers with the information they need to assess students' investigation work.

Other scaffolds can be used by the teacher to model the planning process. Figure 7.7 (the Post-it$^®$ planner) is an example of a planning scaffold that can be made up into a large poster size and used to model the planning process.

The Post-it$^®$ planner starts with clarifying the focus or purpose of the investigation. This might be, for example, discovering what affects the bounce of a ball. Once this has been established the teacher can ask the students what they might change in the investigation, that is, things that might affect bounce. Students might suggest drop height, floor surface, type of ball, etc. Each variable can be written on a Post-it$^®$ note and stuck in the boxes in the planner. The teacher would then ask, 'What will we measure or observe to see what happens to the bounce?' Students might suggest bounce height and the number of times the ball bounces. Again, these variables are written on Post-it$^®$ notes and stuck in the appropriate boxes on the planner. Now that potential variables have been identified, the discussion focuses on which variable will be changed and tested. Students may select drop

Figure 7.6: An investigation planning and reporting scaffold suitable for upper primary students (based on Hackling, 2005, p. 15; reprinted with permission from the Department of Education and Training WA)

Question or prompt	Instructional purpose of the question or prompt
What are you going to investigate?	Students focus on the problem and formulate a question for investigation.
What do you think will happen? Explain why.	Students make a prediction and justify their prediction—this activates prior knowledge for the investigation.
Which variables are you going to: · change? · measure? · keep the same?	Students identify the key variables and decide which ones they will change, measure/observe or keep the same to ensure it is a fair test.
How will you make it a fair test?	Students reflect on their plan and ensure that variables are controlled so that it is a fair test.
What equipment will you need?	Students think about the materials and equipment they will require to conduct the investigation.
What happened? Describe your observations and record your results.	This prompts students to record their measurements and/or observations.
Can your results be presented as a graph?	This prompts students to decide whether it would be worth graphing their data.
What do your results tell you? Are there any relationships, patterns or trends in your results?	These questions prompt students to search for patterns in the data.
Can you explain the relationships, patterns or trends in your results?	These questions prompt students to explain the patterns in their data using science concepts.
Try to use some science ideas to help explain what happened. What did you find out about the problem you investigated? Was the outcome different from your prediction? Explain.	These questions prompt students to summarise their findings as a conclusion and to compare their finding with their prediction. Discrepancies often occur between predictions and findings due to students' alternative frameworks. Such discrepancies may cause students to reflect on their beliefs.

continued over ...

Question or prompt	Instructional purpose of the question or prompt
What difficulties did you experience in doing this investigation?	This question prompts students to reflect on the processes used in the investigation and identify difficulties experienced.
How could you improve this investigation e.g. fairness, accuracy?	This question helps students focus on what they have learned about improving their investigation processes.

Figure 7.7: A Post-it® planning scaffold (based on an original design developed by Goldsworthy & Feasey, 1997, p. 10)

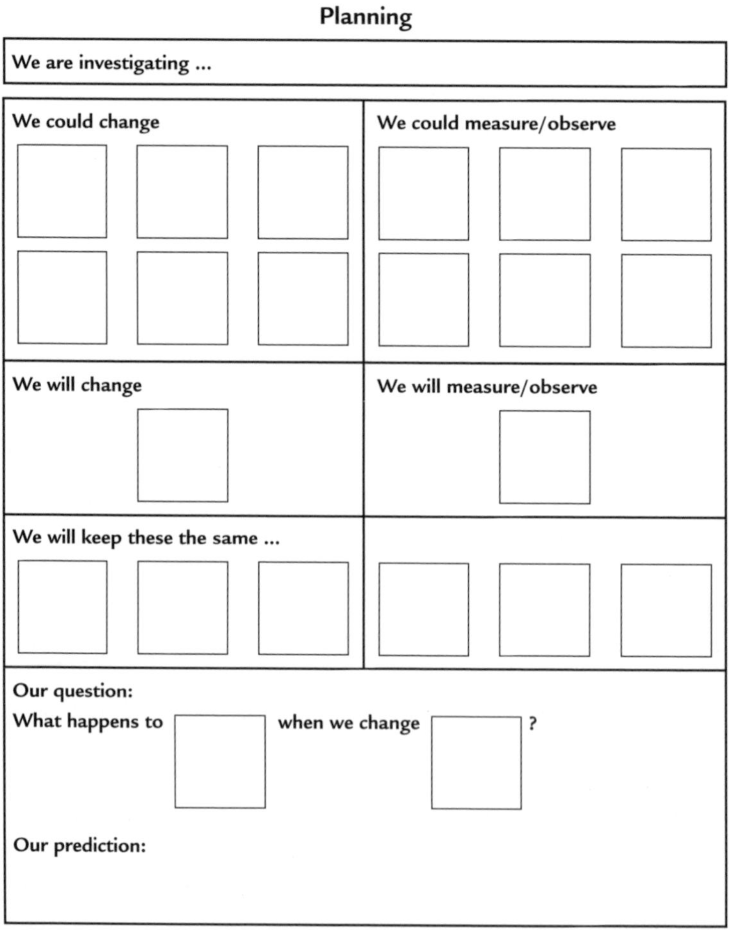

Planning

We are investigating ...

We could change

We could measure/observe

We will change

We will measure/observe

We will keep these the same ...

Our question:
What happens to [ ] when we change [ ] ?

Our prediction:

height and this Post-it® note is then moved down to the next row in the planner. Students next decide which variable will be measured or observed. If students select bounce height, this Post-it® note is then moved down into the next row.

If the investigation is to be a fair test, all other variables that might affect bounce height (floor surface, type of ball) are then moved down into the spaces for variables that must be kept the same. Now that decisions have been made about which variables will be tested (changed) and measured, the research question can be finalised by moving the Post-it® note for the variable that will be measured (bounce height) into the space for *What happens to*, and the Post-it® note for the variable that will be changed (drop height) is moved into the space for *When we change*.

Working with variables

To understand fair testing (controlled experiments) in science, students need to understand the three types of variables and the relationships between them. This is best explained in the context of an investigation of the effect of drop height on the bounce height of a tennis ball. The height to which a ball bounces **depends** on the height from which it is dropped. The bounce height is therefore the **dependent variable** (DV) and the drop height is the **independent variable** (IV). There are a number of other variables that could possibly influence the bounce height, for example, the surface on which the ball bounces and the ball type. These other variables therefore need to be controlled (that is, kept the same) so that we can be sure that any change in bounce height is caused by the changes made to the drop height rather than changes in the surface or ball. These relationships are illustrated in Figure 7.8.

One way of helping students to understand the relationships between variables in a fair test is to use a variables table, an example of which is illustrated in Figure 7.9.

Students can use the mnemonic **C**ows **M**oo **S**oftly to remember to **C**hange something, **M**easure or observe something and to keep everything else the **S**ame to ensure their experiment is a fair test.

A variables grid is another useful scaffold that teachers can use first to help elicit from students the potential variables for investigation and then decide which variable will be changed, measured and kept the same.

Figure 7.8: Relationships between independent, dependent and controlled variables

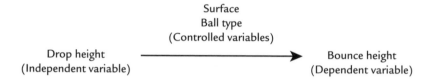

Figure 7.9: An example of a variables table

Question: What happens to the bounce height when we change the drop height?

What I will change	What I will measure	What I will keep the same
The drop height	*The bounce height*	*The type of ball*
		The surface
Independent variable	Dependent variable	Controlled variables

In Figure 7.10, the teacher has first elicited what the investigation is going to find out—in this case the bounce of a ball. This variable, the dependent variable, is placed in the centre of the grid. The teacher then elicits from the class the independent variables that may affect the bounce of the ball. These are written in the surrounding cells of the grid. Once the class has decided which independent variable they will investigate—in this case the drop height—this is the variable they will change. To ensure a fair test, all other variables in the grid must then be kept the same.

Writing research questions

Many students find it difficult to write testable research questions. Primary school students need plenty of experience of working with research questions before being introduced to hypotheses in high school. Research questions are written in the form of a question about the possible relationship between an independent variable and a dependent variable. They can be written in a standard form that can be structured using the following algorithm:

What happens to _____ when we change _____?

        (Dependent variable)   (Independent variable)

Figure 7.10: A variables grid

Material	Colour of ball	Drop height
Type of ball	Bounce height	Temperature of ball
Surface	Size of ball	Force

For example:

What happens to the *distance rolled by a car* when we change the *steepness of a ramp*?

Evaluating the investigation

When students have completed their investigation, it is important for them to reflect on the design of their investigation and their experimental procedure and identify how they could improve their investigation if they did it again. This is the point where students really learn about investigation processes. Some planning and report sheets include questions that specifically address this phase of investigation. For example:

- What difficulties did you experience in doing this investigation?
- How could you improve this investigation, e.g. fairness and accuracy?

Before students conduct their evaluations, either in their groups or individually, it is useful to conduct a class discussion based on a plus-delta chart. This will generate a wide range of ideas. In Figure 7.11 positive aspects of the investigation that should be retained are recorded in the *plus* column, and the things that should be changed are recorded in the *delta* column. (*Delta* is the Greek letter used in science to signify change.)

Figure 7.11: A plus-delta chart

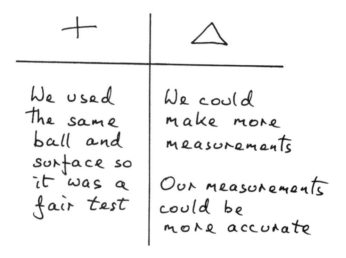

+	△
We used the same ball and surface so it was a fair test	We could make more measurements Our measurements could be more accurate

SCIENTIFIC REPRESENTATIONS OF OBSERVATIONS AND MEASUREMENTS

Norris and Phillips (2003) explain that science literacy entails being able to interpret and construct science texts. As Unsworth (2001) has argued, each discipline has particular ways of representing its ideas, and science has particular conventions when representing observations in the form of diagrams, tables and graphs. These conventions comprise the literacies of science which are explored further in Chapter 10. Students will need explicit instruction in these conventions if they are to construct multimodal texts that communicate the results of their science investigations effectively. The conventions for presenting diagrams, tables and graphs are outlined below.

Diagrams

Science diagrams have a title that explains what the object is. Significant parts of the object are identified with labels which are connected to the parts with arrows. When appropriate, scientific diagrams should include a scale, so the reader can determine how large the object is. In Figure 7.12 the scale is provided as a double-headed arrow with the dimensions marked on it.

Tables

Scientific tables are used to present numerical results such as measurements. Table 7.1 summarises the results from an experiment in which the effect of the drop height on the bounce height of a tennis ball was investigated.

The table has a title which explains what results are in the table and includes the names of the independent and dependent variables. The results for the independent variable are recorded in the left-hand column and the results for the dependent variable are recorded in the right-hand column of the table. The names of the variables are recorded in the column headings, along with the units of measurement.

The purpose of organising results in a table is to help identify the relationship between the variables. In this example, it is important

Figure 7.12: A scientific diagram

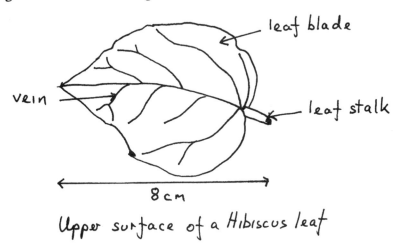

Table 7.1: The effect of drop height on the bounce height of a tennis ball

Drop height (cm)	Bounce height (cm)
150	102
100	68
50	18
20	7

to record the results for the independent variable (drop height) in either increasing or decreasing values; this table shows us that as drop height decreases so does the bounce height.

Graphs

Graphing skills are best developed progressively through a sequence of pictograph, bar graph and line graph. When introducing a new graph type, the teacher must model the construction of that graph type before students attempt to construct their own with careful teacher guidance and support. As they gain experience of the various graph types, students will be able to construct their graphs independently.

The first decision to make when plotting a graph relates to graph type. The teacher must take into consideration the developmental level of the students. Where the results for the independent variable are in categories (e.g. type of ball or surface) and the results for the dependent variable are in the form of continuous data (e.g. numbers, cm, kg, mL) a pictograph (Figure 7.13) or bar graph (Figure 7.14) is plotted. A bar graph is more demanding than a pictograph as it requires the construction of a scale for the dependent variable.

Where the results for both independent and dependent variables are continuous data (e.g. cm, kg, mL) it is conventional to plot a line graph. Line graphs are far more demanding to plot than bar graphs as two scales have to be constructed. A line graph is illustrated in Figure 7.15.

Once the graph type has been selected, the second decision relates to which variable goes on the horizontal axis of the graph. It is a convention that in science graphs, the independent variable is plotted on the horizontal axis.

Figure 7.13: A pictograph showing numbers of students in the class with various eye colours

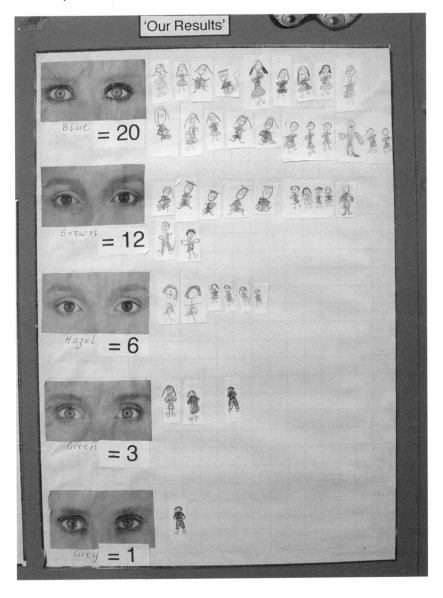

Figure 7.14: A bar graph showing the effect of surface on the bounce height of a tennis ball

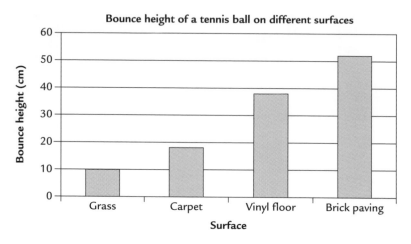

The title of the graph should explain what results are presented in the graph and include the names of the two variables. The independent variable is plotted on the horizontal axis and the dependent variable is plotted on the vertical axis. The variables are labelled on each axis and, where appropriate, the units of measurement are included. The scales for the drop and bounce height measurements start at zero and increase along the axes in equal increments so that scale intervals are equal.

The purpose of constructing graphs is to identify the relationships between the two variables that are being plotted. For example, in the graph of surface against bounce height (Figure 7.14) it can be seen that as the surface gets harder the bounce height increases. In the graph of drop height against bounce height (Figure 7.15), it can be seen that as drop height increases so does bounce height.

Diagrams, tables and graphs are valuable for recording, interpreting and communicating results from science investigations, but they will be effective only if they are correctly presented using appropriate conventions. Students need to be taught these conventions if they are to be successful in constructing and interpreting these text types.

Figure 7.15: A line graph showing the effect of drop height on the bounce height of a tennis ball

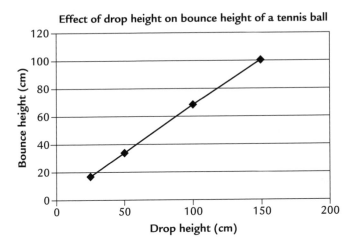

SUMMARY

It is through inquiry and investigation that scientists test ideas and generate new knowledge. By conducting investigations and practising as scientists, students can get a sense of the nature of science and learn skills and understandings at the heart of scientific literacy. Inquiry-oriented science programs actively engage students in learning, stimulate their curiosity, increase their interest in science and develop those critical thinking skills linked to reasoning with scientific processes, ideas and evidence.

ACKNOWLEDGMENT

This chapter incorporates materials from the Australian Academy of Science *Primary Connections* science education publications. We gratefully acknowledge the support of the Australian Academy of Science www.science.org.au, in making its publications available to us for scientific educational use. Australian Academy of Science *Primary*

Connections was funded by the Australian Government Department of Education, Science and Training as a quality teacher initiative under the Australian Government Quality Teacher Programme. www.quality teaching.dest.gov.au.

REFERENCES

Australian Academy of Science. (2005). *Primary connections: Plants in action*. Canberra: Australian Academy of Science.

Goldsworthy, A., & Feasey, R. (1997). *Making sense of primary science investigations*. Hatfield, UK: The Association for Science Education.

Hackling, M.W. (2005). *Working scientifically: Implementing and assessing open investigation work in science* (Rev. ed.). Retrieved on September 28, 2006, from http://www. det.wa.edu.au/education/science/teach/workingscientificallyrevised.pdf.

Hackling, M.W., & Fairbrother, R.W. (1996). Helping students to do open investigations in science. *Australian Science Teachers' Journal, 42*(4), 26–32.

Ministerial Council on Education, Employment, Training and Youth Affairs (MCEETYA). (2005). *National Year 6 science assessment report: 2003*. Retrieved on September 28, 2006, from http://www.mceetya.edu.au/verve/_resources/nat_year6_science_file.pdf.

Norris, S.P., & Phillips, L.M. (2003). How literacy in its fundamental sense is central to scientific literacy. *Science Education, 87*, 224–240.

Unsworth, L. (2001). *Teaching multiliteracies across the curriculum: Changing contexts of text and image in classroom practice*. Buckingham, UK: Open University Press.

CHAPTER 8
RESOURCES FOR PRIMARY SCIENCE CLASSROOMS

Mary Morris and Jennifer Pearson
Edith Cowan University, Western Australia

OUTCOMES

By the end of this chapter you will:
- be familiar with a wide range of resources available to support primary science teachers;
- be able to incorporate a diverse range of resources into primary science learning experiences; and
- be able to develop rich primary science learning experiences incorporating information communication technology (ICT) as a tool.

INTRODUCTION

Teachers in primary schools may find planning and teaching science lessons challenging because of the variety and quantity of resources they may feel they need. Moreover, teachers often are not sure where to find resources or how to obtain, store and make efficient use of them. Some teachers also lack confidence in their ability to teach science because of their own partial understandings of science concepts. Many primary teachers do not complete the 'harder' science subjects in their senior years of secondary education or undertake science content courses at university. The aim of this chapter is to provide primary teachers with information about readily available resources that can enable their students to engage in hands-on activities that will make science memorable.

Students learn about science concepts to help them make sense of the things around them. Using a variety of hands-on materials and equipment can contribute to an improved understanding of these concepts. Students learn best when they construct their explanations for themselves, using their prior knowledge and 'resource-rich' experiences. Thus, the purpose of this chapter is to introduce and discuss resources readily available to primary science teachers including *Primary Connections* (Australian Academy of Science, 2005), *Primary Investigations* (Australian Academy of Science, 1994), *Investigate This!* (Howitt, 2005), *Concept Cartoons in Science Education* (*Concept Cartoons*) (Naylor & Keogh, 2000), consumable and recyclable materials, community and government organisations, ICT and audiovisual materials.

TEXT-BASED RESOURCES

When planning your science program, you need to find out what text-based resources are available within the school. The main criterion for any resource is that it should provide a number of sequential experiences that will develop an in-depth understanding of science concepts. There are two Australian resources that have been developed by teachers for teachers—*Primary Investigations* (Australian Academy of Science, 1994) and *Primary Connections* (Australian Academy of Science, 2005).

Primary Investigations

Primary Investigations has been available in schools throughout Australia since 1994. Its main goal is to sustain children's natural curiosity by encouraging them to explore the world around them and improve their explanations of the phenomena in their world. Its major features include the 5E instructional model (Bybee, 1997), its basis in constructivist learning theory (see Chapter 2), and emphasis on cooperative learning (see Chapter 6), hands-on activities and the use of simple equipment to engage and inform students. Each year level is organised around a major concept and skill through science, technology and environmental aspects linked to hands-on activities. Figure 8.1 summarises the scope and sequence of topics

Figure 8.1: Overview of topics in the *Primary Investigations* resource (Australian Academy of Science, 1994; see acknowledgment at end of chapter)

Topic	Unit 1	Unit 2	Unit 3	Unit 4
Awareness and observation	Introducing awareness of self	Observation	Movement	Space and time
Order and organisation	Introducing organisation	Objects and properties	Materials and structures	Investigating colour
Change and measurement	Introducing change	Comparison and evidence	Tools and machines	Investigating animals
Patterns and prediction	Introducing patterns	Records and data	Constructing and testing	Investigating weather
Systems and analysis	Introducing systems	Interactions and variables	Problems and solutions	Investigating soil
Energy and investigation	Introducing energy	Energy and food chains	Design and efficiency	Investigating astronomy
Balance and decisions	Introducing balance	Ecosystems and resources	Constraints and trade-offs	Investigating materials

within *Primary Investigations*. Within each topic the 5E model provides a number of sequential learning experiences to develop understanding of a concept.

Primary Connections

Primary Connections has been available since 2006 and its purpose is to improve learning outcomes in both science and literacy. It is also structured to improve teachers' knowledge and confidence about science and science teaching. While the 5E model informed the development of this resource, *Primary Connections* is based on an inquiry and investigative approach in which students work from questions to explore their ideas and construct explanations (Hackling & Prain, 2005). The students are given opportunities to represent their developing understandings using a range of texts such as student journals, posters, tables, diagrams, slide shows, web pages and digital photos. Figure 8.2 provides a summary of the topics covered in the resource to date.

Investigate This!

Investigate This! (Howitt, 2005) is one of many resources that are available from science teachers associations around the world. It

Figure 8.2: Overview of topics in the *Primary Connections* resource (Hackling & Prain, 2005, p. 9; see acknowledgment at end of chapter)

Stage	Earth and Beyond	Energy and Change	Life and Living	Natural and Processed Materials
Early Stage 1	Weather in my world	On the move		
Stage 1	Water works	Push-pull		
Stage 2	Spinning in space		Plants in action	
Stage 3			Marvellous micro-organisms	Package it better

focuses on the conceptual area of Natural and Processed Materials and incorporates activities that have been successful in primary classrooms. *Investigate This!* provides an integrated package of learning experiences, including hands-on learning and cooperative group work, and encourages the use of questioning, exploration and inquiry as effective teaching strategies. There is also background information about the scientific concepts covered. The investigation model incorporates skills outcomes as well as providing opportunities for students to further their conceptual understanding and includes six steps for planning science investigations with young children. The six steps are:

1. What are we trying to find out?
2. What do we think will happen?
3. What do we need to do?
4. What did we do?
5. What happened?
6. What would we like to find out next?

Various pedagogical models and assessment strategies are compatible with this resource, such as establishing prior learning through mind maps and cooperative learning strategies such as 'Think Pair Share'. We highly recommend becoming an active member of your state or territory science teachers association. ASTA members have access to a regular journal, an annual science conference and an enormous range of teaching resources.

Concept Cartoons

Constructivist theory (see Chapter 2) suggests that identifying the current level of children's conceptual understanding is very beneficial. One way to do this is to use resources such as *Concept Cartoons* (Keogh & Naylor, 2000). Available as a CD or a book, these cartoons show typical student understandings about a concept/scenario and encourage students to think about their own understanding when they discuss the concept/scenario. This helps the teacher to gather authentic information about their students' alternative conceptions which can then be used to inform the design of learning experiences. The resource also provides ideas for investigations as well as valuable background information on the concepts. An example is reproduced in Chapter 9 (Figure 9.4).

Books

Big Books, non-fiction picture books, posters or black line masters can be used to motivate students, to stimulate questions for inquiry and to support learning of science concepts. Teachers find black line masters useful both for students to record what they are investigating or to verify student knowledge. Teacher development of appropriate worksheets can be time-consuming so even if you need to modify the black line master to suit the needs of your students, you still save time.

RESOURCES WITHIN THE SCHOOL

To enable staff to keep track of their science resources many schools will list such items on a computer database; others may operate a card or book-borrowing system. By talking to other staff members about their science programs you will be able to identify the resources that may need to be shared or alternated between classes. A library support teacher or a Science Coordinator may gather the resources for a teacher upon request. Once you have found what is available it is a good idea to check that the resources work well. Snapshot 8.1 illustrates the importance of this.

Consumable and recyclable materials

These sorts of materials can be purchased through a local supply company or in the local supermarket. There are everyday recyclable materials that are used not only in science, but in technology and art: materials such as cardboard tubes, plastic containers, different-sized cardboard boxes, a variety of plastic lids, fabric scraps and different types of recyclable paper. Early childhood teachers use many of these items in their integrated programs so it is useful for science teachers to liaise with other staff members on popular items. Many teachers put out a list of commonly-used recyclable materials that parents can bring in to school throughout the year. Figure 8.3 shows one way that a school has stored their recyclable materials.

Snapshot 8.1: Magic sorters

Miss Leslie sorted through the collection of magnets the Science Coordinator had given her. She had expected small, round ceramic magnets like the sort you use on the back of fridge magnets, but noticed that these were large bar magnets and some looked a bit old. There wasn't time to change anything so she called the children to the blue mat to begin the lesson of sorting objects.

After an initial discussion about the rules of group work the children were asked to sort a set of objects into groups according to self-selected criteria. After a few minutes, Miss Leslie called them back to the blue mat to share the reasons for the groups they had made.

'Well,' Crystal replied, 'we sorted ours from the ones that we can recycle and ones that we can't.'

'That was a very interesting one,' said Miss Leslie. 'I don't think I would have thought of that interesting one. Thank you, Crystal.'

Miss Leslie congratulated the class on the different ways they'd sorted the objects and went on to describe the next part of the lesson.

'I'm going to come around to your groups now,' said Miss Leslie, 'and I'm going to give you something very special that you're going to use to help you sort all those objects in a different way.'

Miss Leslie handed each group a magnet. In one group, Darcy said that the magnet was 'hard' while Nancy said it was 'metal'. When Miss Leslie 'accidentally' dropped a paper clip onto the magnet it stuck and the children then realised it might be a magnet.

'Ah!' Darcy sang out.

'Oh! What did I do?' Miss Leslie asked.

'It's a magnet,' replied Darcy.

'It's a magnet because they just do this,' said Nancy as she showed the other children how to use the magnet to pick up paper clips. 'They don't pick it up and fall down again. It just picks it up because magnets can pick things up. Oh! It can't pick this paper clip up, but it's metal,' Nancy exclaimed.

'I'll make it,' said Darcy. 'I'll make it go on.' Despite Darcy's efforts he could not get his magnet to pick up the paperclips. Both Darcy and Nancy looked at Miss Leslie, confused. Miss Leslie did not know what had happened and the lesson (to introduce magnets) was not a success.

Miss Leslie was unaware that magnets lose their magnetism over time and had neglected to check that the magnets worked.

Commercially available materials

There is a wide range of commercially available resources to support the teaching and learning of science. These are readily available from department stores, and environment- and science-based education centres, and items can even be provided by parents when they are no longer required at home. These resources include plants, seeds, skeleton models, telescopes, planet models, CDs, posters, model cars and planes, toys, construction equipment such as Lego™, tape measures, magnifying glasses and measuring jugs.

RESOURCES WITHIN THE COMMUNITY

Every community has a wealth of resources that can support a classroom teacher in creating innovative and interesting science programs. There are opportunities to learn from visits to sites such as museums, science centres, zoos, workplaces, universities, landcare sites, commu-

Figure 8.3: Consumables storeroom in a primary school

nity gardens, and much more. Whether your school is located in a small country town, suburb or city will determine the range of facilities available. See Chapter 12 for more information on learning beyond the classroom.

Government departments

Federal, state and territory government departments provide a range of educational materials to promote better understandings of transport options, air quality, water usage, safety with electricity and waste management, to give a few examples. It is worthwhile conducting an internet search to see what is available. In many cases, teaching materials have been created by educators with current knowledge of schools and the curriculum. Types of resources include:

- education kits;
- websites that provide up-to-date and accurate information;
- fact sheets about research;
- professional learning opportunities for teachers;
- educational officers to visit schools;
- funding to support setting up projects;
- competitions for national and state recognition;
- expert advice from scientists; and
- key facilities to visit such as waste water treatment sites.

Local government

The roles and responsibilities of local government are different to those of federal and state governments. At a local level, government officials provide and maintain roads, parks, housing, retail centres and libraries. The resources they offer are more likely to be tied to local issues and history. You need to contact your council or shire for a list of resources. The following organisations may assist with science learning and also link with other learning areas such as society and environment:

- museums of local history;
- local businesses related to science;
- landcare groups;

- rehabilitation sites managed by local government;
- libraries; and
- wildlife rehabilitation centres.

Non-government organisations

Most communities have various non-government organisations that provide information and programs, mainly relating to natural resources issues. The following resources are readily available:

- education kits;
- websites with contact details for programs and support material;
- expert presenters for school visits;
- support networks to link with local community;
- field trips or excursion sites; and
- partnership funding and expert assistance with grant writing.

One example of a useful resource is *Our Wild Plants*, a learning kit produced by Greening Australia (http://www.greeningaustralia. org.au) about the unique flora and fauna of Western Australia. This resource could be readily adapted for teachers in other states. A section of the Greening Australia website titled *Grow us a home* is designed for upper primary school children. The Gould League (http://www.gould.edu.au) runs successful whole-school environmental education programs such as *Sustainable Schools* and *Wastewise Schools* in each state and territory.

These teaching materials need to be considered in terms of the outcomes that you want to achieve. Many schools already possess some of these resources so check the teachers' resource section of your school library or ask your colleagues (who may have the resource you're looking for stored permanently on their shelf). If you are considering taking your class into the community, this topic is discussed in more detail in Chapter 12.

ICT-BASED RESOURCES

ICT provides teachers with a range of innovative and exciting teaching and learning opportunities for science programs. The computer enables you to manage your own teaching programs, lesson plans, assessments, resources and student records electronically. The most common tools used by teachers are Excel™ spreadsheets, word-processing and PowerPoint™ presentations, all of which are available with the standard operating systems on most computers and are essential in the teaching of science. Spreadsheets can be used to record and display tabular data collected during investigations that measure changes in a variable. The students can then re-present their data in a graphical form ready to be placed in their report. Power-Point™ can be used to provide a summary of students' learning. Most schools have access to digital cameras. Digital photos and videos can enhance the quality of PowerPoint™ presentations. A word-processed document can take the place of a journal scrapbook where previously you would have the children write and record their findings. Word-processing packages provide children who have literacy problems with a tool that gives them grammatical and spelling support but allows them to express themselves using a range of templates.

Learning objects—reusable digital resources—are an important ICT resource to support the learning of content in science lessons. The multimedia format of many learning objects combines sound, graphics, text, video and animation in an interactive way. They provide animated models of concepts such as forces and friction showing the microscopic actions that are not visible to the naked eye. As the learning objects become more sophisticated, teachers can access a range of creative and motivational programs. You can source learning objects from a range of websites. Two Australian websites of note are *Teachers on the web* (http://www.teachers.ash.org.au), a website hosted by the Australian Council for Computers in Education, and *The learning federation* (http://thelearningfederation.edu.au), produced by the Curriculum Corporation and funded by the federal government. Davison, Kenny, Johnson and Fielding (2004) discuss the use of learning objects in a primary science classroom and suggest that teachers need to consider whether using a learning object adds value to the learning experience. They also advise that learning objects should 'support, rather than replace hands on activities and authentic inquiry' (2004, p. 9).

Teachers find internet searches useful when looking for information on science topics. While students may enjoy searching for their own information, it is wise to have a list of sites that are relevant and appropriate for primary-aged students. As a pre-service or beginning teacher it is useful to keep a list of favourite websites (as well as a record of any user names and passwords required to access the sites).

AUDIO-VISUAL RESOURCES

Television programs allow children to explore scientific concepts through the use of motivational characters and images. Videos and DVDs allow teachers to be selective about the images they want to use to illustrate a point or provide valuable explanations about concepts. Students can experience the visual impact of watching a machine working, a flowering plant growing from a seed in one minute, animal interactions on a forest floor or a journey through the planets.

Interactive whiteboard technology is becoming more widespread and allows teachers to connect an interactive whiteboard to their computer. In effect, the whiteboard becomes a computer that can be manipulated through touch. The results of a brainstorming session at the beginning of a unit of work can be written on the whiteboard, where they are converted to digital text, saved and stored as an electronic file. A bank of stored templates allows worksheets to be completed in the same manner. You can also search the internet, select information and store it for future use.

Snapshot 8.2 describes how a classroom teacher incorporates ICT skills into her program. The children engage in the science of air quality over three school terms and integrate other learning areas such as Language, Drama, and Society and Environment. Skills developed include use of digital still and video cameras, downloading images, creating and editing movies, writing movie sequences, emailing information, searching websites, storing images, and much more. This highlights the way in which teachers are able to promote a multimedia aspect to science learning.

STORAGE AND MAINTENANCE OF MATERIAL RESOURCES

Teachers have to collect resources for so many learning areas that every square centimetre becomes a multi-use area. Figure 8.3 shows how one school established a highly organised system of storing

Snapshot 8.2: Using technology in the science classroom

In my Year 5 class the computer has become an integral part of daily learning. We have used the computer mainly as a word processor and a research tool but I have extended a range of technology to complement the ICT capabilities of the children.

We have a dedicated weather station in our school grounds as part of the Airwatch program which is offered through state Departments of Environment (e.g. http://environment.nsw.gov.au/airwatch/index/htm). The students are able to download information from the weather station to a computer. From this information, they read the graphs and interpret what is happening with our local climate in relation to rainfall, temperature, and wind speed and direction. The students collate the daily information once a school term and send the results to the local television station and the Department of Environment for their records.

Another aspect of the Airwatch program is assessing the local air quality, which is an issue in most large cities. To assess air quality effectively it was necessary to establish a visual point of reference from the school grounds. A landmark located 25 km away and another 5 km away were needed. Using maps of Perth, the students were able to find the landmarks and determine their distance from the school. Once a month, the students use a digital camera to photograph the landmarks from a set reference point on the school grounds. The photographs are downloaded onto the computer and documented by the students in a PowerPoint™ presentation. Over the course of the year, the students create a monthly photographic diary of changing air quality. The PowerPoint™ diary clearly demonstrates the effects of forest burn-offs, cyclones, smog haze and general weather conditions.

This project was combined with a Travel Smart (http://www.travelsmart.gov.au) program. This program is available in each Australian state and territory. A survey was administered to establish the students' school travel habits which were then collated and analysed using a computer database set up by Millennium Kids (www.millenniumkids.com.au) and the Travel Smart program. The students received instant feedback about their travel habits. On pre-selected dates the students were encouraged to walk to school. A second survey day was carried out and the children completed the same process. The database showed the students how they improved their travel habits and the resulting reduction in greenhouse gases. From this information on the database, students are able to examine how to improve the local air quality and traffic congestion by reducing car usage.

The children then worked with Millennium Kids to create a video looking at local air quality. The class chose to highlight the work they carried out with the surveys and choices they made to get to school. The children wrote a script, prepared costumes and wrote music. The process meant working with a recognised film producer who, with the help of a professional documentary team, filmed the video and involved the class in the editing process. The video demonstrated how the students measured air quality, the reasons why and discussed the weather station. The three-minute video was a suitable culmination of the students' science learning.

materials. The teachers at the school recognised the need to teach children to collect and return resources in an orderly fashion. Parent volunteers can provide assistance, however the ordering of the materials will need to be done by a staff member or Science Coordinator. If there is no central storage area, then teachers should collect only what they need for each school term. You will need to make a list at the beginning of the term and ask parents to collect particular items (e.g. icecream containers). Make sure you let them know when you have enough! Recycling centres, which are often attached to environmental centres, can supply resources that can be collected as needed.

SUMMARY

This chapter recognises that teachers in primary schools often consider science problematic because of the need for extra equipment and resources. In many respects, however, the issue of resources applies to all learning areas, although it is exacerbated in science because teachers are often uncertain of their own science content knowledge. Resources are important to ensure that children have the best opportunities to understand science through hands-on experiences that consolidate their conceptual knowledge and skills. This chapter has suggested a range of resources both within the school and the community. The resources discussed include materials, equipment, personnel and how to find new ideas for your teaching program, as well as the ICT resources that will benefit teaching and learning in primary science lessons.

ACKNOWLEDGMENT

This chapter incorporates materials from the Australian Academy of Science *Primary Connections* science education publications. We gratefully acknowledge the support of the Australian Academy of Science www.science.org.au, in making its publications available to us for scientific educational use. Australian Academy of Science *Primary Connections* was funded by the Australian Government Department of Education, Science and Training as a quality teacher initiative under the Australian Government Quality Teacher Programme. www.quality teaching.dest.gov.au.

REFERENCES

Australian Academy of Science. (1994). *Primary investigations.* Canberra: Australian Academy of Science.

—— (2005). *Primary connections.* Canberra: Australian Academy of Science.

Bybee, R. (1997). *Achieving scientific literacy: From purpose to practices.* Portsmouth, UK: Heinemann.

Davison, J., Kenny, J., Johnson, J., & Fielding, J. (2004). Use learning objects to bring an exciting new dimension to your classroom. *Teaching Science, 50*(1), 6–9.

Hackling, M., & Prain, V. (2005). *Primary connections Stage 2 trial: Research report executive summary.* Canberra: Australian Academy of Science.

Howitt, C. (Ed.). (2005). *Investigate this! Natural and processed materials.* Perth, WA: Science Teachers Association of Western Australia.

Naylor, S., & Keogh, B. (2000). *Concept cartoons in science education.* Cheshire, UK: Millgate House.

CHAPTER 9
USING ASSESSMENT TO IMPROVE TEACHING AND LEARNING IN PRIMARY SCIENCE

Mark W. Hackling
Edith Cowan University, Western Australia

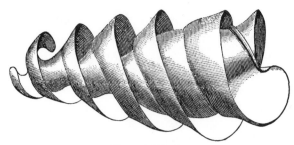

Egg of Shark.

OUTCOMES

By the end of this chapter you will be able to:
- explain the relationships between teaching, learning and assessment;
- use a range of strategies to implement diagnostic, formative and summative forms of assessment; and
- report students' achievement using norm- and standards-referenced frameworks for reporting.

INTRODUCTION

Assessment is the process of gathering and interpreting information about the progress of students' learning. Of all the things teachers do, the one thing that has the greatest effect on students' achievement is effective monitoring of their learning and giving the students feedback about how they can improve. Conducting effective assessment in science requires a rich knowledge of children's learning, science and assessment strategies.

This chapter will address a number of questions related to assessment: What are the relationships between teaching, learning and assessment? How can assessment be used most effectively to improve teaching and learning? How can assessment information be gathered and used to meet feedback and reporting requirements?

TEACHING, LEARNING AND ASSESSMENT

What are the relationships between teaching, learning and assessment? It is worth thinking about your own beliefs about how assessment relates to teaching, learning and reporting to parents before reading this chapter.

The *Professional Standards* of the Australian Science Teachers Association (ASTA) (2002) state that accomplished teachers of science 'use a wide variety of strategies, coherent with learning goals, to monitor and assess students' learning and provide effective feedback' (2002, p. 3). The Victorian Department of Education and Training's *Principles of Learning and Teaching (PoLT)* (2003) include the principle that:

> Assessment practices are an integral part of teaching and learning; and, in learning environments that reflect this principle the teacher:
>
> - designs assessment practices that reflect the full range of learning program objectives
> - ensures that students receive frequent constructive feedback that supports further learning
> - makes assessment criteria explicit
> - uses assessment practices that encourage reflection and self assessment

- uses evidence from assessment to inform planning and teaching.
  (DET, Vic., 2003, p. 1)

These quotes indicate how central good assessment is to effective teaching and learning, and, in particular, the importance of feedback to support further learning.

One of the most profound statements about teaching, learning and assessment was written over 30 years ago by the leading educational psychologist David Ausubel:

> If I had to reduce all of educational psychology to just one principle, I would say this: The most important single factor influencing learning is what the learner already knows. Ascertain this and teach him [sic] accordingly. (Ausubel, 1968, p. iv)

Contemporary constructivist learning theory helps us understand the powerful influence that students' existing conceptions, particularly alternative conceptions, have on their construction of meaning and their learning from classroom activities (see Chapters 2 and 6). Consequently, assessment strategies designed to reveal students' existing conceptions early in the instructional sequence have been included in a number of constructivist-based teaching and learning models. Assessment before and during the instructional sequence can provide information to teachers and students that can be used to improve teaching and learning. Assessment must also occur towards the end of an instructional sequence to determine the extent to which the intended learning outcomes have been achieved. This means that assessment must be embedded within the teaching and learning process and used both to facilitate teaching and learning and for grading and reporting.

PURPOSES OF ASSESSMENT

The purposes of assessment (see Figure 9.1) can be defined in terms of the use to which the assessment information is put. The four main purposes for which assessment is conducted are:

- *Diagnostic* To identify students' prior knowledge and alternative conceptions, so that teaching can be planned to build on existing knowledge and to address students' alternative conceptions.

- *Formative* For providing feedback to teachers and learners so that teaching and learning are improved.
- *Summative* For determining the extent to which students have achieved learning outcomes and/or for certifying achievement and/or for selecting those students who will progress to the next level of education.
- *Evaluative* For evaluating curricula or for accountability purposes, for example, benchmarking the performance of teachers, schools or systems against each other or against defined standards of performance.

Of these four roles for assessment, classroom teachers are only directly involved in diagnostic, formative and summative assessment.

CONSTRUCTIVE ALIGNMENT OF OUTCOMES, LEARNING ACTIVITIES AND ASSESSMENT

John Biggs (1999) has argued that effective teaching and learning require a careful alignment of intended learning outcomes, learning activities and assessment (see Figure 9.2).

This implies that teachers must plan assessment when planning units of work. This will ensure consistency between the expectations of the intended learning outcomes, the learning activities that will enable students to practise the new skills and construct the new understandings required to achieve the outcomes, and the assessment tasks that will gather the information the teacher needs to judge whether the outcomes have been achieved.

Figure 9.1: Purposes of assessment

Purpose of assessment	
Diagnostic:	Assessment for learning
Formative:	Assessment for learning
Summative:	Assessment of learning
Evaluative:	Assessment for quality assurance and accountability

Figure 9.2: Alignment of learning outcomes with activities and assessment

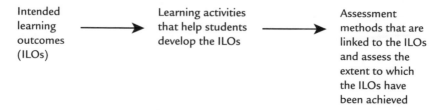

Constructivist alignment is based on two ideas. First, students construct meaning from what they do to learn and, second, the teacher must align the planned learning activities and assessment with the intended learning outcomes. Snapshot 9.1 illustrates constructive alignment. This example demonstrates strong alignment between the expectations of the intended learning outcome, the demands of the assessment task and the opportunities for learning the outcome provided by the learning activities. The criteria of the assessment rubric must be explicit, to ensure that the marking scheme is also aligned to the outcome and the expected level of understanding.

PLANNING FOR THE INTEGRATION OF ASSESSMENT INTO THE TEACHING AND LEARNING PROCESS

One teaching and learning model that exemplifies how diagnostic, formative and summative assessment can be integrated into the teaching and learning process is the *Primary Connections* 5E model described in Chapter 6 (Australian Academy of Science, 2005).

In this model, diagnostic assessment of students' prior knowledge occurs in the *Engage* phase; formative assessment is used to monitor students' developing understandings and give them feedback to extend their learning in the *Explain* phase; and summative assessment is used to evaluate achievement of investigation outcomes in the *Elaborate* phase, and conceptual outcomes in the *Evaluate* phase.

When planning a unit of work, you need to take care to write learning outcomes at a developmentally appropriate level for your students. The state or territory curriculum and assessment frameworks are useful guides for determining the level of the outcomes that

Snapshot 9.1: An example of constructive alignment (based on the *Primary Connections* unit Plants in action, Australian Academy of Science, 2005; see acknowledgment at end of chapter)

Age group: Middle primary
Science topic: Flowering plants

Intended learning outcome
Students will be able to order a set of drawings of stages of the life of a flowering plant into a cycle and describe how one stage develops into the next.

Learning activities
Students germinate broad bean seeds, observe their growth into seedlings and record observations as drawings, measurements and graphs.
   Students study plant flowers and pollination.
   Students cut open fruit and identify seeds within the fruit.
   Students observe examples of dispersal of seeds from fruit.

Assessment task
Students are provided with a handout on which there is a set of drawings depicting the various stages of the life cycle of a bean plant organised in random order. Students are also provided with an A3 sheet of paper, scissors and glue sticks. Students are required to cut out the drawings and paste them on the A3 sheet in a way that illustrates the stages of the life cycle, and then annotate the page explaining how each stage develops into the next.

Assessment rubric

Outcome
Students will be able to order a set of drawings of stages of the life of a flowering plant into a cycle and describe how one stage develops into the next.

Beginning
Students will be able to correctly sequence the life stages but not show the stages linked up into a cycle.

Developing
Students will be able to correctly sequence the life stages and show them as a cycle.

Achieved
Students will be able to correctly sequence the life stages, show them as a cycle and explain how germination, pollination and seed dispersal are processes that link stages of the life cycle.

are appropriate for various phases of learning (see Chapter 4). The National Scientific Literacy Progress Map (MCEETYA, 2005) is another useful guide. Part of the national progress map is illustrated in Figure 9.3.

The outcome in Snapshot 9.1 (students will be able to order a set of drawings of stages of the life of a flowering plant into a cycle and describe how one stage develops into the next) is at Level 3 of the progress map as it requires the student to explain the relationships between individual events that have been experienced, and is appropriate for the middle to upper primary years.

Implementing diagnostic assessment

Diagnostic assessment is conducted at the beginning of a unit or the beginning of a lesson. It is used to determine students' existing knowledge about the topic to be taught, and to identify students' alternative frameworks so that lessons can be planned to build on their prior knowledge and to challenge their alternative conceptions.

Good diagnostic assessment involves the use of probing, open questions that elicit students' understandings of the topic or concept of interest. The simplest way to achieve this is to begin your lesson with a number of open questions to generate a whole-class discussion or brainstorm that will reveal students' existing ideas and beliefs. Once you have ascertained students' existing ideas, you can lead the class into an exploration of the new concept taking an approach that fits with the students' existing level of knowledge.

A number of researchers have developed diagnostic assessment items that have been designed to elicit common misconceptions. In the United Kingdom, Brenda Keogh and Stuart Naylor (2000) devel-

Figure 9.3: Levels of conceptual learning outcomes (This figure is based on a figure in the appendix of the *National Year 6 science assessment report: 2003*, published by MCEETYA, 2005)

Level	Using understandings for describing and explaining natural phenomena, and for interpreting reports	Phase of schooling
1	Describes an aspect or property of an individual object or event that has been experienced or reported.	Junior Primary
2	Describes changes to, differences between or properties of objects or events that have been experienced or reported.	Middle Primary
3	Explains the relationships between individual events that have been experienced or reported and can generalise and apply the rule by predicting future events.	Upper Primary
4	Explains interactions, processes or effects, that have been experienced or reported, in terms of a non-observable property or abstract science concept.	Lower Secondary
5	Explains phenomena, or interprets reports about phenomena, using several abstract scientific concepts.	
6	Explains complex interactions, systems or relationships using several abstract scientific concepts or principles and the relationships between them.	

oped a series of concept cartoons that set a context and pose a question about a natural phenomenon. Three or four children give their explanation about what is happening in the cartoon. Some of the views expressed provide alternative conceptions while others provide a scientific explanation. You can use the cartoons to stimulate discussion between students by asking them, 'What do YOU think?'

The *Science Education Assessment Resources* (SEAR) (Commonwealth of Australia, 2005) online bank of assessment resources is another good source of diagnostic assessment items. Some of these items use cartoons (see Figure 9.5) to set the context and pose questions to elicit students' conceptions. The items also provide detailed information

Figure 9.4: A concept cartoon (Naylor & Keogh, 2000; reproduced with permission)

What do YOU think?

for the teacher explaining the correct scientific explanation, the alternative conceptions and what steps the teacher can take to address the students' alternative conceptions.

Diagnostic items can be used to stimulate small-group discussions, which make both teacher and students aware of the range of conceptions existing in the class. This can lead to investigations to test the various views held by the students. A useful strategy to keep track of students' conceptions through a unit of work is to maintain a class TWLH chart. Figure 9.6 shows a TWLH chart which was developed by the *Primary Connections* Project (Australian Academy of Science, 2005, p. 7).

Initial conceptions that have been diagnosed early in the unit are recorded in the first column of the chart. Questions students wish to

Figure 9.5: SEAR diagnostic assessment item 4LL115 (adapted from *Science Education Assessment Resources* (*SEAR*), 2005, copyright Commonwealth of Australia, reproduced by permission)

Is the fire alive?
Explain why the fire is/is not alive.
What should you look for to tell if something is living?

Figure 9.6: A TWLH chart (Australian Academy of Science, 2005, p. 7; see acknowledgment at end of chapter)

What we **think** we know.	What we **want** to learn.	What we **learned**.	**How** we know.

investigate during the unit are recorded in the second column and the new ideas developed through the unit are recorded in the third column. The fourth column records evidence gathered in the unit that supports each of the new ideas recorded in the second column. As this evidence is reviewed it will be evident that some of the ideas recorded in the first column at the beginning of the unit are contradicted by the evidence, and should now be discarded and deleted from the chart. This process helps students continually monitor ideas and evidence and the relationships between them, and build an understanding of the way science develops evidence-based conclusions.

Implementing formative assessment

Formative assessment is used to provide feedback to teachers and learners so that teaching and learning are improved. The monitoring of learning and giving of feedback occurs in two main settings. The first is when the teacher is working with small groups or with individual students. The teacher notices evidence of the students' level of thinking from their written work or explanations and then responds with a question or comment that makes the students think more deeply about the idea. The second setting is when teachers are marking students' work and give feedback to students as written comments on the work.

Wiliam (1998) argues that for assessment to be truly formative five conditions must be met:

- a mechanism must exist for determining students' current level of achievement;
- a desired level of achievement, above the current level, is identified;
- the two levels are compared so that a gap is identified;
- the teacher provides prompts, scaffolding or support that informs students about *how* to close the gap; and
- the learner uses the information to close the gap.

Formative assessment can be structured around profiles of learning outcomes that are set out in levels and describe a progression in learning. Figure 9.7 illustrates a progression in learning in the processes of interpreting evidence and drawing conclusions, and is based on the National Scientific Literacy Progress Map.

Figure 9.7: A progress map of outcomes relating to interpreting evidence (This figure is based on Table A1.1 in the *National Year 6 science assessment report: 2003*, published by MCEETYA, 2005)

Level	Interpreting evidence and drawing conclusions
1	Describes what happened.
2	Makes comparisons between objects or events observed.
3	Displays data as tables or bar graphs, identifies and summarises patterns in science data. Applies the rule by extrapolating or predicting.
4	Calculates averages from repeat trials or replicates, plots line graphs where appropriate. Conclusions summarise and explain the patterns in the data.

A teacher who understands this progression in learning is empowered to make sense of students' attempts at interpreting evidence and to provide useful feedback that can help them make more sophisticated interpretations. This is illustrated in Snapshot 9.2. This example of effective formative assessment is based on an understanding of progression in learning and carefully structured feedback.

Sadler (1989) argues that for students to take control of their learning and production of quality academic work, they must:

- possess a concept of the standard being aimed for;
- compare the current level of performance with the standard; and
- engage in appropriate action which leads to closure of the gap.

Teachers must provide their students with information about the standard of work that is expected. This can be achieved in standards-referenced forms of assessment where learning outcomes are described in profiles of levels of achievement. Levels of performance can be communicated to students through descriptive statements, checklists, rubrics and/or exemplars.

One of the most effective ways of helping students to understand the outcomes they are working towards is to involve them in self- and peer-assessment. Students who work in pairs, using a checklist or

Snapshot 9.2: An example of formative assessment in action

A teacher working with Year 2 students who are investigating factors that influence the bounce of a tennis ball may interact with a child who says, 'When I dropped the ball on the floor it bounced up to my knee'. Recognising this as a Level 1 response and that the next level requires students to make comparisons, the teacher might say, 'Go and try it on the lawn and see what happens'. Once the child has tried bouncing the ball on the lawn, the teacher might say, 'What did you find out?', to which the child may respond with 'It did not bounce as high on the lawn as on the floor'. In this example, by providing appropriate prompts and questions the teacher has supported the child to move from Level 1 to Level 2 thinking in relation to interpreting evidence and drawing conclusions.

rubric to comment on each other's work, can compare their own and their partner's work objectively to the standards described in the rubric. This allows them to develop an understanding of the expected standards and what aspects of their work need to be improved.

The main factor that determines the effectiveness of formative assessment is the quality of the feedback and how it is given. Feedback must be task-involving rather than ego-involving, that is, it should direct attention to the task rather than to the self or ego. Feedback in the form of congratulatory comments such as 'Good work' or marks such as 3/10 or 9/10 are ego-involving. To maximise achievement gains, feedback should be task-involving and include comments that describe what the student is doing well, what needs to be improved and gives some guidance on how to improve the work, for example, 'Your diagram shows all the parts of the flower and you have used labels to identify the parts. What could you add to your diagram to show the reader how big the flower is?'

Implementing summative assessment

Summative assessment is used towards the end of a unit, term, semester or year to determine the extent to which students have achieved the learning outcomes. Teachers use this information for grading or determining the level of achievement, and reporting to parents. To determine their students' grades for the semester, teachers typically use a range of evidence—classroom observations, students' oral reports, students' science journals, reports of investigations and other work samples which may be assembled into a portfolio. Using a range of sources of evidence provides a more valid, reliable and equit-able assessment of students' achievement in science.

Methods of collecting and interpreting the evidence for your summative assessment and reporting must satisfy the assessment framework or frameworks required by your school, educational sector, state or territory and by the Australian government.

Standards-referenced and norm-referenced assessment frameworks

Standards-referenced assessment involves reporting achievement in relation to standards of performance. One way of doing this is to report achievement for each subject/learning area in five bands, which may be designated A, B, C, D and E and/or with descriptors such as Excellent Achievement, Good Achievement, Satisfactory Achievement, Limited Achievement and Cause for Concern.

Norm-referenced assessment involves reporting achievement in relation to the student's cohort of peers, that is, all students in that year group. A student's achievement may be reported in terms of the performance of the student in a given subject/learning area as being, for example, in the first, second, third or fourth quartile of students. The first quartile includes the top 25 per cent of students who have the highest aggregate scores attained over the reporting period.

Planning assessments

As argued earlier in this chapter, you need to plan your assessments when you are planning units of work, so that the assessments align with the intended learning outcomes. You also need to take into account the purposes for which the assessment information will be used. It is most likely that you will use your assessment information to provide feedback to students about their learning and to parents. Given that reporting to parents may include both norm- and standards-referenced forms of reporting, you need to take these three main uses into account when planning and conducting your assessments.

You should plan to assess your students' achievement in terms of the intended learning outcomes. Assessment should sample the full range of outcomes and, at least, include both investigation and conceptual outcomes of science. You need to plan for assessments that will be comparable to those used by other teachers teaching the same year cohort at their school. Assessments need to be of a comparable standard and need to assess the same outcomes in similar ways, particularly when using a norm-referenced framework. It is a good idea, therefore, for you to plan units of work and assessments as a collaborative task with colleagues teaching at that year level.

We suggest that teachers should design rubrics to assess students' achievement of outcomes at a number of levels or standards in relation to those outcomes. Experience suggests that three levels is an optimal number. Once you have made initial assessments of outcomes using a rubric, you can use this information to give feedback to students and to report to parents about students' achievement in science in both norm- and standards-referenced forms. Each state and territory provides guidelines for teachers about how to generate grades for reporting based on their assessments.

Best practice in reporting

Reporting to parents is most effective when a number of strategies are used. These should include:

- sending home information on a progressive basis so that parents/guardians are alerted to concerns early enough in the school year to take action on those concerns;
- meeting with parents/guardians and the child with a portfolio of work samples so that performance can be illustrated in meaningful and concrete contexts;
- including information about achievement in relation to standards and in relation to their peer group in students' report cards;
- explaining the grading system in report cards in plain English free of educational jargon;
- including information in report cards on work effort, social and emotional development in addition to academic achievement; and
- explaining what the child is doing well, what needs to be improved, and how that improvement can be made.

Most education departments and curriculum and assessment authorities provide guidelines to their teachers on assessment and reporting requirements. These are worth reviewing before planning your assessment and reporting strategies and materials.

SUMMARY

As the primary purpose of schooling is to improve learning, assessment must focus on providing feedback to both teachers and students that can be used for improving teaching and learning. To do this assessment must be embedded within the teaching–learning process, rather than tacked on the end of the learning sequence as a measurement of the extent of learning. Constructive alignment of intended learning outcomes, learning activities and assessment will ensure coherence and integrity in the teaching–learning–assessment processes. Careful planning of assessment is needed to ensure that the assessment information collected will satisfy the purposes of providing feedback to both learners and teachers, and of reporting to parents in norm- and standards-referenced forms.

ACKNOWLEDGMENT

This chapter incorporates materials from the Australian Academy of Science *Primary Connections* science education publications. We gratefully acknowledge the support of the Australian Academy of Science www.science.org.au, in making its publications available to us for scientific educational use. Australian Academy of Science *Primary Connections* was funded by the Australian Government Department of Education, Science and Training as a quality teacher initiative under the Australian Government Quality Teacher Programme. www.quality teaching.dest.gov.au.

REFERENCES

Australian Academy of Science. (2005). *Primary connections: Plants in action*. Canberra: Australian Academy of Science.

Australian Science Teachers Association (ASTA). (2002). *National professional standards for highly accomplished teachers of science*. Retrieved on May 5, 2006, from http://www.asta.edu.au/home/whatsnew/profstandards.

Ausubel, D.P. (1968). *Educational psychology: A cognitive view*. New York: Holt, Reinhart & Winston.

Biggs, J. (1999). *Teaching for quality learning at university*. Buckingham, UK: Society for Research into Higher Education (SRHE) and Open University Press.

Commonwealth of Australia. (2005). *Science education assessment resources*. Retrieved on May 5, 2006, from http://cms.curriculum.edu.au/sear/.

Department of Education and Training, Victoria. (2003). *The principles of learning and teaching*, p. 12. Retrieved May 5, 2006, from http://www.sofweb.vic.edu.au/pedagogy/plt/principles.htm.

Ministerial Council on Education, Employment, Training and Youth Affairs (MCEETYA). (2005). *National Year 6 science assessment report: 2003*. Retrieved

September 29, 2006, from http://www.mceetya.edu.au/verve/_resources/nat_year6_science_file.pdf.

Naylor, S., & Keogh, B. (2000). *Concept cartoons in science education*. Cheshire, UK: Millgate House.

Sadler, D.R. (1989). Formative assessment and the design of instructional systems. *Instructional Science, 18*, 119–144.

Wiliam, D. (1998). Enculturing learners into communities of practice: Raising achievement through classroom assessment. Paper presented at the European Conference on Educational Research, Ljubjana, Slovenia. Retrieved January 12, 2004, from http://www.kcl.ac.uk/depsta/education/hpages/dwiliam.html.

PART III
EXTENDING THE ART OF TEACHING PRIMARY SCIENCE

CHAPTER 10

LEARNING THE LITERACIES OF SCIENCE

Vaughan Prain
La Trobe University, Victoria

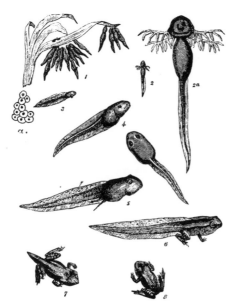

Life-stages of Frog

OUTCOMES

By the end of this chapter you will:
- understand the nature of the literacies of science;
- be able to plan, implement and evaluate students' learning of the literacies of science; and
- understand the developmental aspect of the literacies of science across the years of primary schooling.

INTRODUCTION

There is growing agreement among science educators that learning science is not just about understanding concepts but is also about learning the subject's particular literacies. These include the subject-specific vocabulary of topics and the distinctive ways in which science concepts, processes and findings are represented through combining verbal, visual and numerical languages (Moline, 1995; Gee, 2004; Lemke, 2004; Australian Academy of Science, 2005). To learn the literacies of science, children must draw on the literacies they already know and use—their everyday language practices of talking, listening, reading, writing and drawing, interacting with electronic technologies and expressing ideas through gestures and role-play. This chapter focuses on effective use of literacies in learning science.

Every science topic in primary school requires students to engage with the subject-specific ways of representing science. For example, when students are learning about the life cycle of plants in Years 4 and 5, they are expected to know how to interpret and construct the kind of diagram presented in Figure 10.1, which is from the *Primary Connections* Plants in action unit (Australian Academy of Science, 2005).

Figure 10.1: Cross-section of a flower (Australian Academy of Science, 2005; see acknowledgment at end of chapter)

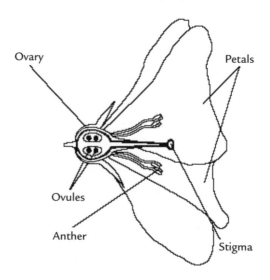

While this diagram looks relatively easy to understand, it implies various interpretive challenges for students in learning the particular literacies of this subject. Students need to recognise that it is a cross-section labelled diagram of a flower, where lines are used to indicate parts of the structure, in this case specifying the scientific terms for these parts. They may know about a cross-section from life experiences of apples or oranges being cut in half, but they need to recognise that cross-sections are used to show scientific explanations because they focus on key elements. They also need to know that lines may serve diverse purposes in diagrams, and can be used to indicate directional flow, a sequence of actions, or processes in a chain of events, such as the stages of a plant's life cycle. Students need to recognise and use these varied yet specific purposes as part of acquiring science literacy.

According to Unsworth (2001), every visual representation in science has recognisable structures that indicate (a) the experience, object or event referred to; (b) how viewers or readers are expected to interact with the text; and (c) how the text is organised on the page or screen. In Figure 10.1, the object or flower is intended to be an example of flowers in general, and thus this image is meant to represent the parts common to all flowers. The flower is the whole focus of the image, but other images may depict only part of the whole, and zoom in on a key area to focus on sub-functions within the whole. In terms of the style of representation, the image in Figure 10.1 is a simplified line drawing of its object, but scientists also use more complex diagrams and photographs to represent evidence, findings and explanations. Students need to learn how to interpret and use these different kinds of representations.

The image in Figure 10.1 could be one in a sequence depicting the formation of fruit, and students need to learn how the passage of time is represented in scientific diagrams. This can be indicated by numbered or captioned images, cartoon boxes, or re-representations of a slightly modified image of the subject. While other texts, such as comics and recipes, use these ways of depicting the lapse of time, students need to understand the use of these conventions in representing scientific accounts of natural processes. Images that convey scientific information also usually avoid an orientation where the viewer looks up or down at the subject, but instead present a level frontal perspective, as in the case of the flower image in Figure 10.1. This perspective is intended to give a more objective or reliable viewpoint to the object. Another way in which images are presented as objective is through depicting hidden parts as visible, as in cutaways of internal parts or cross-sections, again evident in this image. In other words, a single image can entail a wide range of interpretive demands on students learning the new literacies of science. The next section considers more broadly the particular literacy products and practices entailed in constructing scientific representations.

THE LITERACY PRODUCTS OF SCIENCE

In engaging with science learning in primary school, students need to learn how to interpret and construct scientific ideas, processes, methods and findings through the integrated use of factual texts, drawing, diagrams, tables and graphs. These texts can be used to explore and clarify understanding as well as to indicate a more resolved understanding. As noted by Lemke (2004) and many others, this combination of verbal, visual and mathematical modes is an essential feature of scientific explanation. In observing and measuring changes to phenomena and developing explanatory accounts of these changes, primary students are fundamentally engaged with acquiring the many facets of scientific literacy. The following points consider key features of the main text types encompassing the literacies of science, and focus on their possible applications to the topic of plant growth.

Factual texts

While factual texts are common to many school subjects such as art and history, they are crucial for presenting concepts or findings on science topics. Such texts are usually multimodal, that is, they incorporate verbal and visual language. They also include various textual organisers to present content clearly—titles, labels, highlighted key terms, diagrams, maps and photographs. A **science journal** may include students' written accounts of speculative questions and answers on a science topic or inquiry, along with their observations, investigations and reflections arising from science lessons. It might contain a sequence of chronological entries, each with the date on which the entry was made, and a heading relating to these observations. These entries may be multimodal and include evidence such as digital photographs of plant growth, captions, labelled diagrams, tables of results and graphs of changes to plants' shoots. Students might also use their journals to reflect on their success in learning about particular topics. As noted by Gee (2004) and Unsworth (2001), while students might use first-person oral and written narratives to describe investigations in junior primary classrooms, by the upper years of primary school they are expected to have learned how to use the passive voice in writing on science topics. This latter style seeks to place less emphasis on the role of students as agents in investigations. For example, 'We measured plant growth each day' is an example of a first-person account, whereas 'Plant growth was measured each day' is an example of the use of passive voice.

Diagrams

Various kinds of diagrams are used to report science findings. A labelled diagram aims to capture the appearance and features of an object or process more clearly than is possible through verbal language alone. A diagram has a title, a scale to indicate the subject's size, and captions and lines to identify key features. In the topic of plants, students typically are expected to learn how to make labelled diagrams of plant parts, plant growth, cross-section drawings of flowers and the life cycle of flowering plants. They are expected to learn about all processes in a plant's life cycle and generate appropriate explanatory captions for each stage.

Data charts

A data chart is a text for organising information and consists of a title and columns in which information is presented under appropriate headings. For example, as in Figure 10.2, data charts can be used to investigate conditions that affect plant growth, where students record over several days the effects on seed growth of exposure to different conditions, such as different growth mediums or access to contrasting amounts of water and light.

Flow chart

A flow chart is a graphic way to describe a sequence of events, forces, stages or phases in a process. Such sequences can be represented in a

Figure 10.2: An example of a student's data chart recording conditions affecting plant growth

Day	Plant with water	Plant with oil
Monday	6 cm high 7 cm wide no flowers some green leaves some yellow leaves 50 mL water	7 cm high 7 cm wide one flower three small yellow leaves no green leaves 25 mL oil
Wednesday	6 cm high 7 cm wide no flowers two dead leaves no wilting some new growth 50 mL water	7 cm high 8 cm wide flower dead one dead leaf some new growth 25 mL oil
Monday	6 cm high 7 cm wide no flowers many dead leaves no new growth 50 mL water	5 cm high 10 cm wide dead leaves 25 mL oil
Wednesday	8 cm high 7 cm wide no flowers some dead leaves new green leaves 50 mL water	3 cm high 10 cm wide no flowers some dead leaves other leaves dark green and purple 25 mL oil
Friday	8 cm high 7 cm wide some new growth one purple leaf some new green leaves soil moist	3 cm high 9 cm wide purple leaves plant sagging soil greasy

linear fashion along one line, with arrows indicating the order in which the sequence occurs, or represented through a circular pattern. Flow charts such as Figure 10.3 are an enabling way for students to attempt to represent their understanding of a plant's life cycle.

Figure 10.3: Flow chart of the life cycle of a plant

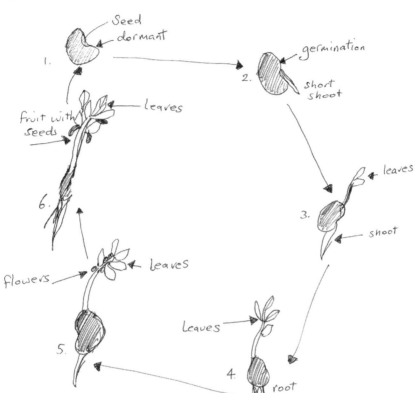

Tables and graphs

A table is an organisation of words, numbers or items into columns and lines in order to show relations between the items. A graph is a symbolic representation using lines or systems of lines (bar graphs) to symbolise variations in the occurrence of items.

Scientific evidence is often summarised and represented in tables and graphs, where both types of text present data to clarify relationships between variables. The columns of a table and the axis of a graph are used to present data about each set of variables. Both tables and graphs need to show the units used to measure data, so that the reader can understand the effect of the variables.

In interpreting and constructing these texts, students need to learn the conventions governing which variables are located on columns of tables and on the axes of graphs. The variable in a table that is changed (the independent variable) is placed in the left-hand column while the variable affected by the change and usually measured (the dependent variable) appears in the right-hand column of the table. In graphs, the independent variable is placed on the horizontal axis, while the dependent variable is located on the vertical axis.

In both tables and graphs the data make clear how the variables relate to one another. For example, in a table where students record the effect of access to different amounts of water on plant growth (one variable) in one column of a table, with the effect of different amounts of sunlight in the next column (another variable), the students can decide whether there is any pattern of interaction between them. Students need to understand that organising data into tables and graphs provides a way to identify patterns which can guide further findings and the development of justifiable claims. In this way, learning the literacies of science is not just about understanding the conventions of these texts, but also how to interpret them effectively to build understandings and findings. Similarly, students need to seek to explain patterns in bar heights in a bar graph (see Figure 10.4) and the slope of the line in a line graph (see Figure 10.5) as indicating links between the variables. This will lead to further interpretive work.

GENERAL LITERACY TEXTS

There are many general literacy processes and products that support learning of the literacies of science, and the following examples are intended as indicative only.

Procedural texts are common in many school subjects, and in students' conversations, and aim to explain *how* to do something. They may include lists of ingredients or materials required to complete a task, or provide a sequence of directions and instructions. On the topic of plant growth, students could set out instructions for other students on how to conduct an investigation into plant life, how to produce seedlings from seeds and how to identify the conditions that affect plant growth. They could construct oral, or print- or

Figure 10.4: A student's bar graph of daily temperatures

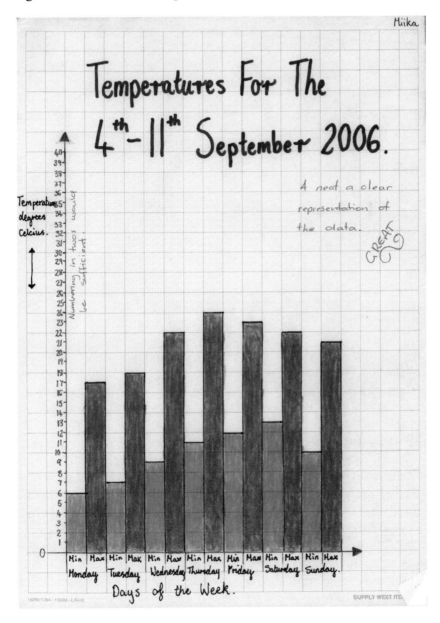

Figure 10.5: A student's line graph of a plant's growth

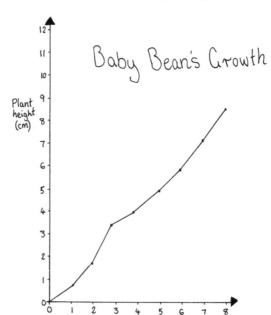

computer-based texts. They could also make oral presentations on what they have learned about plants, including how to draw parts of plants accurately.

A **recount text** usually describes past events or personal experiences, focusing on *when*, *where* and *why* the events took place. Such texts typically start by setting a context or establishing background information, then report on what happened, sometimes including the narrator's attitude towards the experience. This everyday textual experience can be easily adapted to science learning. Students can write accounts of their predictions, observations and investigations in learning about the life cycle of plants.

With the guidance of their teacher, students can create a **word wall** indicating their initial and developing sense of relevant vocabulary for understanding and representing the topic of plant growth. An initial word wall could be constructed by asking students to categorise different objects, including seeds, into 'living' and 'non-living'.

A different word wall could be constructed around student's own questions in relation to different aspects of plant growth. Students could also write **summaries** or **mind maps** of different stages in a sequence of lessons about plants. They might also produce maps, models or video presentations of the plants and flowers in their immediate environment to identify common features.

Students could conduct **interviews** with members of their community who are involved in plant and crop propagation, such as plant nursery workers, farmers and gardeners. Such interviews could be structured, where the students design in advance the questions they wish to ask, or semi-structured, where the students follow up set questions with further queries depending on the issues raised. Students could also write individual, paired or group letters seeking information about plants from appropriate experts in their community. Some students could make group poster-supported oral presentations on the topic, or write or present a narrative text to demonstrate their understanding of key concepts associated with the life cycle of plants. The topic of plant growth also provides considerable scope for relevant guided discussion after reading and viewing experiences to consolidate learning, for example, a class could watch and discuss David Attenborough's (2003) video on flowering and pollination.

PLANNING AND IMPLEMENTING A PROGRAM TO SUPPORT LEARNING THE LITERACIES OF SCIENCE

Effective program development for the learning of science literacies requires teachers to be confident about their knowledge of key concepts in the science topics, to have a clear sense of the learning goals for lesson sequences and to have a rich diversity of teaching and learning strategies and resources to support effective student engagement with this learning. Students need many opportunities to participate in a representationally-rich learning environment if they are to learn how to interpret and construct the kind of texts that represent scientific habits of mind and reasoning processes, and records and findings of their inquiries. They need to be challenged to make, modify, link and justify many different kinds of representations as they clarify key concepts, methods and processes in any science topic. This learning will be enhanced where students experience a diverse range of learning contexts, including small-group or paired discussion to whole-class investigation, and resources such as group posters, traditional and interactive whiteboards, computer-based programs, role-playing, and digital cameras to represent scientific ideas.

Primary Connections (Australian Academy of Science, 2005) is an example of a professional development program for teachers that aims to address this broad range of issues. It identifies the following teaching and learning strategies to enable effective links between science and literacy learning in primary classrooms:

- the teacher models how she or he would construct a text such as a graph or a table to represent findings from an inquiry;
- the teacher guides whole-class discussion on the function of specific textual features in the kinds of texts the students are expected to interpret and construct;
- the teacher and the whole class jointly co-construct a science text relevant to the topic using a poster or traditional or interactive whiteboard;
- the teacher supports individual students and student groups to construct representations of their current understanding of the topic;
- the teacher guides student reading, viewing and listening to promote text interpretation;
- students participate in cooperative learning groups where they are assigned roles to support focused engagement with inquiry, and report back to the whole class;
- students as individuals and in groups construct texts incorporating multimodal representations of topics;
- student groups discuss their understanding of the purpose of key textual features in their own and others' science texts;
- students make oral, written and electronic presentations to a range of different audiences and readerships to demonstrate their learning of a topic;
- students reflect orally and in writing on what helps them to learn to engage with science topics effectively;
- students role-play and perform their understandings; and
- students are given opportunities to initiate their own inquiry into a science topic of interest to them.

These strategies imply that teachers need to provide an appropriate mix of guided and more independent learning experiences for students as they develop an understanding of how scientific inquiry and findings are represented. During the primary school years students need to be given a range of opportunities for interpreting and constructing various kinds of tables, graphs and diagrams as tools for scientific thinking and as records of their science activities.

EVALUATING LEARNING THE LITERACIES OF SCIENCE

While teachers need explicit criteria for evaluating students' literacy products in science, it is also important that students develop their own knowledge of these criteria. They need to be able to monitor the effectiveness of their own learning. The following points elaborate criteria for interpreting and constructing multimodal science texts for Years 6 and 7. Multimodal texts that link visual, verbal and mathematical information are becoming more prevalent with increasing student access to computer-based technology such as the internet, CDs and DVDs. Examples of multimodal texts include Rio Tinto's *Big Science Online* (http://www.asi.edu.au) and the *Science Learning Objects* produced by the Learning Federation (http://www.thelearning federation.edu.au). Many of the *Science Learning Objects* provide models, simulations and demonstrations of scientific concepts and practices. They provide teachers and students with experiences that are not available universally because they require expensive equipment, are unsafe or occur over extended periods of time. While the following points are presented as resolved, the students, with their teacher's guidance, could consider developing their own criteria for what makes a science text interesting, effective and reliable. As noted already, multimodal texts are essential to learning science in that they are the main way in which scientific explanations are represented in the science community. While Lemke (2004) has noted that new technologies and new web-based reports continue to modify the structure and language of how scientific activity is represented, students must necessarily work with multimodal texts if they are to develop science literacy.

INTERPRETING AND CONSTRUCTING MULTIMODAL SCIENCE TEXTS FOR YEARS 6 AND 7

Multimodal science texts link verbal, visual, mathematical and aural language to convey themes, concepts, relationships and explanations. For students to interpret these texts successfully they need to be able to understand the main components of the text, including the conventions used, and the global meanings generated by the linkage of these parts. To construct these texts themselves, students need to know how to integrate different modes in ways consistent with science conventions in order to make a coherent and persuasive whole.

Students need to understand how these texts are organised, the rules used for layout across the page or screen, and ways to focus viewer attention. Some common conventions are:

- familiar material is usually put on the left-hand side of the page, with new or more demanding content on the right;
- layout usually parallels the direction of print, going from top to bottom;
- parts may be separated or joined using various framing devices, including arrows, line borders, blank space, or overlapping or superimposed images. Strong frames signal sharply separated parts; and
- emphasis in layout is achieved by variations in size, colour and position of particular images on the page, by animation effects and by framing of parts.

Questions to guide students' interpretation of multimodal science texts

The following sets of questions are intended to encourage students in Years 6 and 7 to be critical readers and viewers of multimodal science texts. While not all questions are appropriate for any particular text they are meant to provide broad scaffolding for text evaluation. Some of these questions could be used or adapted to guide discussion with younger students.

Questions to ask when students first look at science texts

What first catches your eye in this text? Why?

In what order did you view/read the parts of the text? Why was this the case?

Do you think other readers/viewers will follow a similar pattern? Why/why not?

Questions that focus on features of science texts

How many parts are there in this text? What is each part trying to achieve? Is the purpose clear?

Which part is most important? How can you tell?

Are all the parts linked together? Why/why not?

How effective for you is the mix of different modes (verbal, visual, mathematical, aural languages) used in this text?

What other texts are similar to this one? Is this text as clear as these other texts?

What claims or explanations are being made?

Questions about science texts that encourage further inquiry

Who produced this text? How do you know?

Who is expected to read this text? How do you know?

What claims and evidence are presented in this text? Can you trust them? Why/why not?

Who or what is shown in this text? Who or what has been left out? Why?

Could further important claims or evidence be added to this text?

Does this text give a fair account of its topic? Why/why not? What further information do you need to answer this question?

Are different interpretations of the topic of this text possible?

Do other scientific texts present opposite or different findings on this topic? Which text should you trust and why? Should you wait for more evidence?

How might this layout be re-constructed to make its topic clearer or more convincing?

What does the maker of this text want you to know or do?

How well did this text answer what you expected?

Questions to guide students' construction of multimodal texts

Students also need to develop a framework for guiding their own text production, and the following sets of questions provide some scaffolding.

Questions to ask when students are starting their own text production

What is the main point of your topic?
What descriptions, explanations or claims will you make?
What evidence will you use to support your claims?
What visual and other resources do you have to make the topic clear and interesting?
How will you catch the attention of your reader/viewer?
In what order do you want the parts of your text read or viewed? Why?
How will you guide your reader/viewer in engaging with your text?

Questions to ask about the features and structure of science texts

How many parts should your text have? How will you make the purposes of each part clear?
Which is the most important part? How will you show this?
Are all the parts of your text linked together to make up a whole?
Have you planned an effective mix of different modes (verbal, visual, mathematical, aural language) for this text?

Questions to encourage students to check their science texts

What claims and evidence do you want to present in your text? How can you be sure they are reliable?
Who or what is shown in your text? Has anything important been left out?
Could further important claims or evidence be added?
Does your text cater for different readers/viewers with different background knowledge of the topic?
What do you want your readers/viewers to know or do after interpreting your text?
How well did your text meet your expectations?

Table 10.1: Developmental map of literacies of science products from the *Draft literacy focuses progress map* (Hackling & Prain, 2005)

Stage	Science journal	Factual texts	Diagrams	Tables	Graphs
Emerging	Teacher-modelled whole-class science journal	First-person student oral presentation/ demonstration	Teacher-captioned student drawing	Teacher-constructed whole-class table	Teacher-scaffolded whole-class pictograph
1	Teacher-modelled whole-class science journal Individual student science journal	First-person student-written recounts including illustrations Teacher-guided whole-class poster Individual role-play	Student-captioned drawing using some conventions such as arrows	Student-recorded data in teacher-supplied table	Individual student pictographs
2	Individual student science journal	Procedural texts Summaries Posters Reports incorporating multimodal representations	Student-drawn cross-section with labelled parts Mind maps Data charts	Teacher-supported individual student-constructed simple tables	Individual student bar and column graphs
3	Individual student science journal with increasing focus on multimodal representation and reflection	Investigation reports incorporating third-person, passive voice construction Oral presentation supported by 2-D and 3-D representations such as posters, Powerpoint™, models and demonstrations	Student scale drawings from different perspectives Cutaways Flow charts Concept maps	Individual student tables	Graphs including teacher-supported individual student simple line graphs

DEVELOPMENTAL MAP OF LITERACIES OF SCIENCE PRODUCTS

To optimise opportunities for students to develop an understanding of the literacies of science over the years of primary schooling, it is useful to have a broader developmental picture of what types of texts students should engage with at different year levels, and in what ways. Clearly some topics are more suited to particular types of text. For example, it makes sense for students to work on tables and flow charts when recording their investigations relating to plant growth. However, there is value in students engaging with increasingly challenging texts of broadly the same kind as they progress through primary school. Table 10.1 presents an indicative framework for such a progression across four broad stages. This progress map was developed for the *Primary Connections* Project (Hackling & Prain, 2005). The development of students' use of science journals, factual text, diagrams, tables and graphs is described.

SUMMARY

This chapter has argued that learning the literacies of science is crucial in enabling students to engage effectively in learning science. These literacies, like students' everyday literacies, are powerful resources for learning and key tools for thinking. They also entail products, processes and practices that indicate what students have learned. As teachers, we need to create a representationally-rich environment where students use both everyday and scientific literacies to engage successfully in scientific ways of thinking, acting and constructing explanations.

ACKNOWLEDGMENT

This chapter incorporates materials from the Australian Academy of Science *Primary Connections* science education publications. We gratefully acknowledge the support of the Australian Academy of Science www.science.org.au, in making its publications available to us for scientific educational use. Australian Academy of Science *Primary*

Connections was funded by the Australian Government Department of Education, Science and Training as a quality teacher initiative under the Australian Government Quality Teacher Programme. www.quality teaching.dest.gov.au.

REFERENCES

Attenborough, D. (Producer). (2003). *The Private Life of Plants. Episode 3: Flowering.* London: BBC Worldwide Ltd.

Australian Academy of Science. (2005). *Primary connections: Plants in action.* Canberra: Australian Academy of Science.

Gee, J.P. (2004). Language in the science classroom: Academic social languages as the heart of school-based literacy. In E.W. Saul (Ed.), *Crossing borders in literacy and science instruction: Perspectives on theory and practice* (pp. 13–32). Newark, DE: International Reading Association and National Science Teachers Association.

Hackling, M.H., & Prain, V.P. (2005). *Primary connections stage 2 trial: Research report.* Canberra: Australian Academy of Science.

Kress, G. (2003). Genres and the multimodal production of 'scientificness'. In C. Jewitt & G. Kress (Eds.), *Multimodal literacy* (pp. 182–190). New York: Peter Lang.

Lemke, J. (2004). The literacies of science. In E.W. Saul (Ed.), *Crossing borders in literacy and science instruction: Perspectives on theory and practice* (pp. 33–47). Newark DE: International Reading Association and National Science Teachers Association.

Moline, S. (1995). *I see what you mean: Children at work with visual information.* Portland, ME: Stenhouse Publishers.

Unsworth, L. (2001). *Teaching multiliteracies across the curriculum: Changing contexts of text and image in classroom practice.* Buckingham, UK: Open University Press.

CHAPTER 11
DESIGN PROJECTS THAT INTEGRATE SCIENCE AND TECHNOLOGY

Stephen Norton
Griffith University, Queensland
Stephen Ritchie and Ian Ginns
Queensland University of Technology, Queensland

OUTCOMES

By the end of this chapter you will be able to:
- describe the subtle differences between science and technology;
- justify the integration of science and technology in design activities; and
- begin to plan integrated science and technology units of work.

INTRODUCTION

What is the difference between science and technology? In the sixteenth century, science and technology were virtually indistinguishable. Leonardo da Vinci, for example, drew sketches of a helicopter and explained that if the helical surfaces were made 'airtight with starch [and] rotated with speed that the said screw bored through the air and climbs high' (Leishman, 2000, p. 4). It is difficult to determine if this is a technologist or a scientist talking because da Vinci refers to both the material properties and the causal effect of a rotating spiral upon air currents. As we will see in this chapter, the differences between science and technology are subtle.

These subtle differences can be illustrated with reference to the design process articulated in *Primary Investigations* (Australian Academy of Science, 1994)—a set of resources commonly used to support the teaching of science in primary schools. For example, in the sixth student book that focuses on Energy and Change, there is a section that describes the design process in terms of continuing cycles of Design, Make and Appraise (DMA). Design refers to the artefact you intend to make and includes making decisions about the problem you want to solve, being clear on what the invention is intended to do and deciding on materials. The design stage may involve brainstorming and drawing plans. Making is the construction phase, and students appraise their inventions by testing them and deciding how good they are. The design activities in this *Primary Investigations* student book are intended to link design with the study of energy efficiency. In this exemplar, there is a critical difference between science and technology. In science, design is secondary to inquiry, and the focus is on applying science to explain the artefact or developing a better understanding of scientific phenomena through construction and appraisal. In technology, in contrast, the creativity of a design of a useful artefact is central. Thus, although the two disciplines have much in common, their purpose is different. In the past this has resulted in a craft orientation to the teaching of technology and a more academic approach to the teaching of science.

Technology practice embodies the actions of:

- investigation (identifying the problem and gathering information and data);

- ideation (planning and designing);
- production (creating and making); and
- evaluation (testing, judging and refining).

It is tempting to view these actions as being linear in nature and plan accordingly. In practice, however, students may move in almost random ways between the actions such that the overall process appears to be cyclic, iterative or even recursive.

The content of recent curricula in technology has been divided up into subunits, or strands. For example, in the Queensland *Technology Years 1 to 10 Syllabus* (Queensland Studies Authority, 2003) these strands include Information, Materials and Systems. Information is generally seen as the purposeful organisation and communicating of data and since the advent of information communication technologies (ICTs) this is frequently seen as 'operations with computers'. While a computer is a wonderful tool to collect, process, manipulate and present data, it is but one tool that is available for technology in the Information strand. Others include drawing protocols, multimedia tools including video and audio, art protocols and the construction of flow charts.

The Materials strand focuses on the observable and useable technological properties of materials (e.g. strength, flexibility, plasticity), and transforming and manipulating materials. This can be compared with science programs' focus on understanding the nature of materials and their properties. The nature of a material (e.g. its molecular structure) may help us to explain, or account for, its technological properties.

The Systems strand of technology refers to the study and development of a combination of components that work together to achieve a specific purpose and can also be described in broad terms of relationships between inputs, processes and outputs. Flow charts, models and schematic diagrams can be used to represent systems. Similarly, in science, interactive components can be seen in the study of nature (e.g. biological systems) and interacting components of the universe in general. These components frequently are represented using the same tools as technology systems.

So we can now put forward a tentative answer to the original question about the differences between science and technology as defined by emerging curricula. Simply put, science focuses more on understanding the nature of the world, while technology involves the application of information, materials and systems to solve problems of a practical and applied nature.

WHY TRY TO TEACH SCIENCE AND TECHNOLOGY TOGETHER?

If there are such subtle differences between science and technology, why is it worth integrating these subjects in classroom activities? The first answer is time. One of the main reasons primary teachers give for not teaching science is the lack of time available in the school curriculum to cover all subjects or learning areas in depth. Thus, integrating science and technology in design activities can save time. The second answer is that the motivational potential of technology practice can be used to stimulate student interest in science (Zubrowski, 2002), particularly for those children who enjoy making artefacts for a purpose (e.g. Norton, McRobbie & Ginns, 2004). A third reason is that integrated activities help students to learn with greater understanding by enabling them to investigate and grapple with the multiple relationships that exist between components of knowledge and skills (Murdoch & Hornsby, 2003). Integrated, authentic and student-centred learning is also thought to assist in the development of thinking structures including reflection and metacognition (De Bono, 2005), and to capitalise on different learning styles as described by Gardner's theories of multiple intelligences (Gardner, 1993). Finally, this form of learning is thought to help students to develop creativity. Creativity has been defined as the production of novel thought solutions, or products based on experience and knowledge within an environment that includes risk taking (Cole, Sugioka & Yamagata-Lynch, 1999). What is clear from the above summary is that novel, innovative or creative solutions tend to come from adding a new layer to past experience and knowledge.

ILLUSTRATING THE INTEGRATION OF SCIENCE AND TECHNOLOGY

Students bring to the classroom diverse and imaginative ideas for technology design projects. Tapping into children's interests and creativity

Snapshot 11.1: Design project—the Giant Drop

After their visit to Dreamworld three girls, one each in Years 5, 6 and 7, elected to construct a model of the Giant Drop ride back in the classroom. They were able to use digital photographs, video footage, notes and sketches of the ride compiled during their visit as information for initial planning. However, this information appeared to dominate their thinking, thus leading to them replicating the ride rather than generating their own design ideas. Their attention early in the design process focused on what materials they would need in order to make the tower. The students resolved numerous design and production problems including producing a suitable tower, a mechanism to guide the falling passenger cage, a braking mechanism to slow the cage as it neared the end of its drop, and mechanics to winch the passenger cage back to the top. They finally used a cardboard cylinder supported with angle brackets to make the tower, dowelling rods and plastic tubing to guide the falling cage, with sticky tape as a brake. An electric motor was connected to a winch in order to pull the passenger cage to the top of the tower. However, the girls could not manufacture a gear box to slow the winch despite having access to gear box components purchased from a local electronics shop. This failure was due to difficulties the girls experienced in trying to mesh the correct size cogs, so they decided to use a simple hand winch that they constructed from Lego™ pieces.

In describing their ride, the students used the following science-related terms: pulley; motor too strong (powerful); going around and around fast (rpm); slide down easier (friction); push (force); it's falling faster and faster (acceleration due to gravity); time (minutes and seconds); fast speed (velocity in metres per second and kilometres per hour); going faster (acceleration); units of measure (centimetres and metres); standing up (stability). The classroom teacher introduced the concepts of energy, both potential and kinetic, and the relationships between these forms of energy and momentum. To afford students and teacher the opportunity to discuss the meaning of science concepts in the context of the design activity, the group explored mathematical topics associated with space, measurement and number including proportional thinking. Space and measurement concepts were particularly important in the planning phase where scale diagrams were constructed.

To assist the process of connecting the activity to science outcomes the teacher provided students with access to text resources on how mechanical devices work. The use of materials and the control of the ride were linked to technology outcomes.

Students were unanimous in their belief that the integrated project was educationally valuable. For example, Suzie reported that: 'There was a lot of measurements in our project. I liked building and stuff. It was hard, but we had to try and once it was figured out it was easy'. Another student, Heather, said: 'Maths and science has made [it] easier by building our giant drop and now we understand how it works'.

Figure 11.1: Design project—the Giant Drop. Note the two dowelling rods to guide the passenger cage and a small manual winch behind the tower. The poster contains explanations of key concepts associated with the project. The students are presenting to classmates and invited parents.

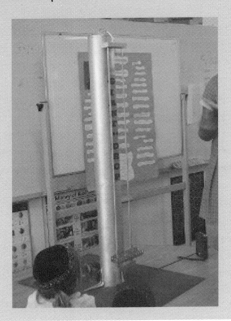

Snapshot 11.2: Design project—toy cars

Two girls and a boy in Year 3 participated in this design project to build a model of a Dreamworld ride of a vintage car on tracks. The children drew rough sketches of toy cars and made a list of components they would need. The complexity of constructing a car to run on tracks was daunting so the teacher organised the selection and purchase of a suitable kit from an electronics store. With limited assistance from the teacher, the children followed instructions to cut and glue the car together. However, the car's wide wheelbase meant that it could not turn on a guiding track constructed out of a rubber tube glued to a 1.3 m by 1.7 m piece of plywood. After much discussion between the children and the teacher, the children decided to use a Meccano™ set to construct a motor and gearbox. However, the plans were not clear and the nuts and screws so small that these children did not have the manual dexterity to build the artefact. They then decided to use Lego™ pieces to make a smaller toy car (Figure 11.2).

Figure 11.2: The composite Lego™ car with an electric motor and pulley drive. The rubber tubing guiding track was glued to a plywood base.

Figure 11.3: Completed toy car project, including chart of definitions, initial kit car and track environment.

The following is a summary of students' descriptions of their construction.

Toby: We made it out of Lego™. We superglued the motor onto the car so it would not fall off.

Brittiny: The motor was an electric motor. It has batteries. Wire was on the battery. The energy came from the battery. We had to switch the wires around because it was going the wrong way.

| Elly: | The motor made the pulley go, which made the wheels go. We had to take the tyre off, the wheel was too big and it would not stay on the track. |
| Brittiny: | It took seven seconds to go around the track. Our car had energy, it could go. The band could stretch like an elastic band. |

can be an exciting way to start a topic suitable for developing both science and technology concepts. Snapshots 11.1 and 11.2 describe two rich design projects that stemmed from a school excursion to Queensland's Dreamworld fun park. The snapshots illustrate how teachers can capitalise on students' interests and creativity, as well as achieve science and technology outcomes. They also provide evidence of several key pedagogical issues that may affect attempts to integrate science and technology. By inviting her students to build a model of a Dreamworld ride, the teacher's intention was to involve her 12 students (composite Year 1–7 class) in projects that would challenge them to think about science and technology. (Dreamworld has several exciting rides that fall, spin, coast, and so on.) During the day excursion, students first experienced the physical sensations of riding the Giant Drop (a passenger cage is dropped from a height of about 80 metres and accelerates due to gravity until its descent is slowed and stopped by huge electromagnets), and enjoyed rides in vintage cars on tracks. The snapshots of the two design projects also illustrate the big ideas associated with Energy and Change (science) and with design processes (technology).

The teacher had the students present their artefacts to their parents and the wider community. Their presentations included explaining the underpinning technology, science and mathematics both verbally and visually on a chart. This presentation process appeared to help the students make connections between their creations and explicit learning outcomes as well as motivating them to complete and refine their artefacts by a set deadline.

Snapshot 11.3 illustrates the use of Lego Robotics™ in design activities where science and technology learning outcomes can be integrated.

Snapshot 11.3: Working with robotics

Lego Robotics™ is a sophisticated learning environment that combines two central aspects: first, the physical construction of machines and, second, the provision of programming opportunities using a simple, icon-based programming language. Potentially, Lego Robotics™ offers students an opportunity to engage in engineering at an advanced level and to avoid many of the pitfalls that can occur when working with materials that are hard to manipulate.

In one robotics challenge, Year 7 students were asked to design a robot to push cans out of a circle. Figure 11.4 shows a can-pushing robot. The students used two motors, gears, touch sensors, wheels of various sizes and different problem-solving strategies to succeed in their endeavours. This challenge was a rich opportunity for technology and science learning associated with the physical construction of a robot.

The second aspect of robotics is that of programming. The icon-based programming language of Labview (National Instruments, 2006) enabled students to program their robots to search for cans, respond to touch, push the cans out of a circle and turn back into the circle after sensing the black border of the circle. The program contains splits, loops, lands, repeats and instructions to motors to move forward or reverse at various power settings for various times. Clearly, the act of programming was a rich opportunity to transform information, a central aspect of technology practice. In addition, data-collecting probes, or sensors, (e.g. pH, sound, temperature, rotation, velocity, acceleration, light) can be used, all of which can be manipulated and represented with Lego™ software, thus adding a sophisticated 'information' component to the design challenges. Programming in conjunction with probes can be used as a way for students to visualise control in systems, a concept that enables us to explain how numerous artefacts such as air conditioners and automatic gear boxes work.

Figure 11.4: A can-pushing robot with touch feelers that locate the cans. Note the use of a jockey wheel at the rear of the robot to provide an element of variability in the search pattern.

Of course, teachers can integrate science and technology learning outcomes without having to rely on commercially-produced resources. Everyday household materials can be used to challenge students' design skills. Cardboard boxes and tubes can provide resources for the design of machines while matchsticks and even spaghetti can feature in the design briefs for bridges. However, when providing materials such as spaghetti to build bridges the teacher might reflect on the authenticity of constraining student creativity with the use of such materials. Roth, Tobin and Ritchie (2001) detail case studies of learning science through projects in the design of machines, buildings and bridges.

LEARNING FROM DESIGN ACTIVITIES

Several important implications can be drawn from the design projects discussed in Snapshots 11.1, 11.2 and 11.3.

- Students can construct authentic design solutions to particular design briefs involving the union of technology and science.
- Occasionally, students experienced difficulty in constructing the artefacts that they had planned. This difficulty in translating creative ideas into a functioning reality was linked to the students' familiarity with the material resources and skilled use of the

available tools. Teachers should be aware of the potential for unnecessary frustration when design briefs are not matched appropriately with skill levels. Increasing the demands progressively through related design briefs might provide the sort of scaffolding needed to help students realise positive outcomes from their design activities.

- With respect to using kits of various sorts teachers would be well advised to trial the kits before their use in class. In the case of Lego Robotics™ kits, teacher professional development, and support from community members or from a local high school teacher with expertise in robotics are highly recommended.

- It was observed that throughout all the activities teachers had to encourage students to articulate the underpinning science and technology principles. In planning the integrated teaching of science and technology, therefore, teachers should invest time in identifying the overlaps between the conceptual outcomes and make them explicit to the students. Many design projects offer the opportunity to integrate beyond technology and science to include mathematics, study of society, and literacy, for example.

ASSESSMENT OF INTEGRATED SCIENCE AND TECHNOLOGY PROJECTS

A fundamental principle of assessment is whether the student can communicate his or her understanding in meaningful ways. Various forms of assessment can be used. Teachers can observe work in progress and keep written notes, checklists, photographs, audio records of conversations between students working in their groups and conversations between the teacher and individual students. (Note the value and richness of students' comments in the snapshots.) Work samples (e.g. drawings, sketches, flow charts, artefacts, explanations and discussions) can be analysed for evidence of the processes and knowledge associated with the artefact's construction and performance. The performance of an artefact can be judged against set criteria. Teachers can negotiate these criteria with the students as part of the action of evaluation in technology practice. Determining the extent to which students demonstrate autonomy and abstraction might help teachers to assess the designing process. Finally, students' understanding of science and technology knowledge can be assessed by determining the extent to which they interconnect the concepts embedded in the science and technology of their project and transfer them to new design projects in new contexts.

SUMMARY

There is considerable overlap in the ways of working scientifically and working technologically. In other words, the two ways of investigating and manipulating the environment complement each other. However, the expression of outcomes in emerging curriculum documents indicates nuances that set science and technology apart and teachers need to be aware of these. When teaching in an integrated science and technology environment it is easy to slip into the habit of assuming that students who have constructed a technological artefact have achieved science outcomes. One way of countering this is to plan to make both learning outcomes explicit and to create a classroom environment in which students are supported to demonstrate success in meeting both science and technology outcomes.

With this in mind, the following points may guide your planning of integrated science and technology.

- Choose tasks that provide authentic learning opportunities for the students and that are rich in science and technology concepts.
- Analyse the task(s) for the key science and technology outcomes that are prerequisite, embedded or associated with the projects. Construct a table or concept map of outcomes that shows how they relate to each other.
- Decide if and what aspects of other subjects or learning areas you wish to integrate and identify their key outcomes. Language, Studies of Society and Environment (SOSE), Mathematics and Art can be integrated as well, especially in planning and in presenting the process and product to an audience.
- Identify materials, cost and expertise that might be needed to conduct the projects. Consider inviting participation from community members with special skills (e.g. during production).
- Provide a stimulus to engage students in investigating and ideation of their project. You may have to provide them with

scaffolding to draw plans, construct flow charts, write lists and develop criteria for their designs. Scaffold students through their investigation and ideation. Remember all the actions of technology practice are iterative.

- Take advantage of 'point of need' opportunities (teachable moments) to help students see the links between the activity and the science and technology outcomes. Decide which concepts will be taught in parallel (i.e. separate from the production) and which will be taught within the technology practice activity (i.e. during production).
- Decide what techniques you are going to use to assess the outcomes. Remember, you want the students to be able to make the critical concepts explicit. Some form of presentation will probably form part of your assessment package (e.g. during production and evaluation). Include students' self-evaluation of their learning.
- Reflect on the success of the project. Make notes for future integrated tasks. Remember you have to consider a multitude of issues including students' cognitive and affective outcomes, scaffolding activities, time constraints and costs.

REFERENCES

Australian Academy of Science. (1994). *Primary investigations. Teacher resource book: Energy and change.* Canberra: Australian Academy of Science.

Cole, D., Sugioka, H., & Yamagata-Lynch, L. (1999). Supportive classroom environments for creativity in higher education. *Journal of Creative Behavior, 33*(4), 277–293.

De Bono, E. (2005). *Thinking course: Powerful tools for transforming your thinking.* Bath, UK: CPI.

Gardner, H. (1993). *Multiple intelligences: The theory in practice.* New York, NY: Basic Books.

Leishman, J. (2000). *A history of helicopter flight.* Retrieved November 10, 2005, from http://www.enae.umd.edu/AGRC/Aero/history.html.

Murdoch, K., & Hornsby, D. (2003). *Planning curriculum connections: Whole-school planning for integrated curriculum.* South Yarra, Vic.: Eleanor Curtain Publishing.

National Instruments. (2006). *Labview: Graphical programming for engineers and scientists.* Retrieved May 12, 2006, from http://www.ni.com/company/robolab_labview.htm.

Norton, S.J., McRobbie, C.J., & Ginns, I.S. (2004). Student approaches to design in a robotics challenge. In H. Middleton, M. Pavlova & D. Roebuck (Eds.), *Proceedings of the 3rd Biennial International Conference on Technology Education Research Vol. 3* (pp. 26–35). Brisbane, Qld: Griffith University.

Queensland Studies Authority. (2003). *Technology Years 1 to 10 Syllabus.* Retrieved March 22, 2007, from http://www.qsa.qld.edu.au.

Roth, W-M., Tobin, K., & Ritchie, S. (2001). *Re/constructing elementary science.* New York: Peter Lang.

Zubrowski, B. (2002). Integrating science into design technology projects: Using a standard model in the design process. *Journal of Technology Education, 13*(2), 48–67.

CHAPTER 12

TEACHING AND LEARNING SCIENCE AND TECHNOLOGY BEYOND THE CLASSROOM

Janette Griffin and Peter Aubusson
University of Technology, Sydney, New South Wales

OUTCOMES

By the end of this chapter you will be able to:
- develop learning programs that integrate learning beyond the classroom with learning in the classroom;
- use a framework that provides students with a clear purpose for visiting sites outside the classroom; and
- encourage student choice and ownership of their learning by providing stimulating and worthwhile experiences.

INTRODUCTION

When we reminisce about our own school days, some of the most vivid memories will be of experiences that occurred outside the classroom, perhaps on a camp or an excursion. A fabulous way to captivate and motivate primary-aged children about science and technology is to tap into the natural motivation and excitement generated by experiences outside the classroom. Primary-aged children also need structured practice at learning from and understanding the wide world outside their everyday experiences. This chapter provides advice about teaching and learning science and technology that takes place beyond the classroom. It addresses appropriate ways in which you can integrate day-to-day learning within the classroom with carefully planned, novel visits to sites outside the classroom.

Studies of children's natural learning with family or friends have helped us understand how children can learn beyond the classroom (see, for example, Hein, 1998). The informal learning that children experience with family and friends can be full and wondrous, but it also tends to be haphazard. Classroom learning is usually more targeted and structured. In this chapter we will show how the informality that characterises learning at sites such as museums, zoos, bushland, parks and local shopping centres can be used to link school-based and site-based teaching to create seamless, authentic learning. (We are using the term 'site' to encompass the varied places that teachers may take their students.)

Children first arrive at school with about five years of accumulated experiences and understanding of their world. They have learned with and from their families, friends and many others; in the back yard, at playgrounds; on visits to places formal and informal. This outside learning continues after starting school.

As teachers, we recognise students' learning experiences outside the classroom, not only by taking into consideration their prior knowledge, but also by celebrating these experiences and their influence on the children's attitude to learning, interests and curiosity. We should provide opportunities for children to continue to learn outside the classroom, not in a separate and disconnected way, but as an integral part of schooling. We can take children into the real world of the playground to explore leaves and plants, insects and birds, shadows and buildings (Lucas, 1997). We can take children to the local park, bushland or streets to explore complex social, natural and built environments. We can explore more structured environments such as zoos, field study sites, museums, gardens and heritage sites. Figure 12.1 shows Year 6 students collecting water samples at a local lake where they conducted an intensive study of the water quality.

WHAT WE HAVE LEARNED ABOUT LEARNING BEYOND THE CLASSROOM

Research on learning in settings beyond the classroom (Griffin, 2004) has revealed a lost opportunity, with many teachers failing to embrace the different learning experiences intrinsic to sites. One reason for this is the tendency to transfer modes of teaching/learning from formal classroom settings to informal learning settings. Out-of-class-room sites are not, and should not be treated as, classrooms. A solution lies in a more natural, informal mode of learning suited to the first-hand experiences that abound there—a mode of learning revealed by studies of site visits by friendship and family groups.

Figure 12.1: Students collecting water samples at a lake near their school

Family group studies at sites reveal rich social interactions and extensive informal learning (Falk & Dierking, 2000). Curiosity takes hold, questions are raised and answers sought by children. Extended conversations, including sharing ideas and opinions, are normal. Children choose things that interest them to investigate, and choose how long they spend doing it. There is authentic engagement with selected exhibits and locations that promotes deep understanding. The learning is open-ended, often raising issues and puzzles for further investigation.

Research has demonstrated that successful school visits to sites are learning-oriented and stimulate open-ended exploration. Unfortunately, studies of many school groups revealed visits characterised by control, with children strictly directed where to go and what to do, and allowed limited choice of their own. Worksheets provided to the students managed learning with minimal input from teachers (Griffin & Symington, 1997). This closed approach yields superficial learning with students' records of information demonstrating superficial treatment of exhibits and locations. Learning was not fuelled by curiosity and direct experience but by written responses to questions designed by site experts with no knowledge of the class, what learning had gone before or what would come after. Visits were isolated events rather than part of an integrated learning program. It is perhaps unsurprising that many site visits have been less fruitful than they could have been. Interviews with teachers reveal that they often feel uncertain when taking students on excursions (Griffin, 2004). In the absence of an appropriate pedagogy, many teachers' behaviours reflected the way they had been taken on excursions when they were school students. Feeling insecure or concerned for their school's reputation, teachers often hid behind a screen of formality and control.

In short, school group learning beyond the classroom has often been task-oriented with documentation an end in itself. The many productive, self-directed features identified in family visits suggest an alternative way forward.

APPROPRIATE PEDAGOGY: *SMILES*

Research in Australia during the 1990s led to the development of a teaching/learning framework called *SMILES (School–Museum Integrated*

Learning Experiences for Students) (Griffin, 1996). (The word 'museum' is used here as a generic term for any site, indoor or out, visited by a class group.) It has been tested and used by teachers, schools and institutions in Australia and internationally. The key to effective school site visits is not just the strategies teachers use at the site but the suite of learning provided before, during and after the visit. Primary-aged children achieve extensive learning through site visits when they:

- know the purpose of the visit;
- have some control over their learning;
- have choice about what, where and how they learn; and
- work with small groups.

Learning beyond the classroom must be integrated with learning in the classroom. This does not merely mean that the topic should be addressed in lessons before and after a site visit but that an appropriate learning approach, one designed to marry the best features of learning in formal and informal settings, is applied. *SMILES* does this.

The theoretical basis of *SMILES* is social constructivist (see Chapter 2). More specifically, *SMILES* is based on the interactive teaching approach (Biddulph & Osborne, 1984; see Chapter 6 for more information). It requires teachers to select the learning experiences based on the students' needs, anticipated uses, or interest. It involves self-directed pacing, both physically (e.g. the speed the learner moves through a site), and also mentally (e.g. when to stop and engage intently and when to scan superficially). The learning relies on children raising their own questions, which permits them to drive the direction and depth of their investigations.

While concentrating on the appropriateness of learning opportunities beyond the classroom, there is no denying that learning environments outside the classroom present particular challenges. Teachers are, justifiably, concerned about taking children to unfamiliar places but a key concern should be to integrate the management and learning issues. Too much emphasis on control and supervision can smother learning. Conversely, real learning is promoted when management is well-planned in advance. Successful management strategies include:

- planning the visit with the children;
- informing and preparing parent helpers;
- becoming familiar with the site; and
- planning for safety, travel and eating.

Because of the need to integrate learning at the site with learning in the class, *SMILES* is organised into three phases; before, during and after the visit. Here the description is detailed for clarity. The depth to which you apply *SMILES* will depend on the age and the experience of your students as well as your learning goals. We have provided most detail on preparation before the visit as this is critical to successful learning, however all three components are essential.

AT SCHOOL BEFORE THE VISIT

Determining areas of inquiry

Introduce the class to the topic and help the children formulate areas of inquiry or questions. Students could brainstorm, draw pictures, answer a quiz, or draw a mind map of ideas about the topic and questions they want to explore. Developing general areas of investigation may be more successful than highly specific questions. For example, 'What kinds of plants grow in the park?' is more likely to be a fruitful question than 'Does spinifex grow here?' Gather and share a full set of class questions. Form groups of children with common interests. The groups can begin investigating their part of the topic through class activities, library or internet research, personal interviews and other lines of inquiry. Ask the children to record questions that arise as they are investigating their aspect of the class topic. These are the learners' questions that guide further learning at school and on the site visit. See Figure 12.2 for an example of questions raised by a group of Year 4 students before a zoo visit.

Snapshot 12.1 provides a more elaborate example from a Year 6 class that was involved in two visits to a zoo. One group of students asked a question about what meerkats (an African animal) do all day and used research prior to the zoo visits, as well as structured observations during the zoo visits, to help them answer their question.

Figure 12.2: Examples of questions raised by Year 4 students before a zoo visit

---

Is a reptile an animal? How can we tell?

Is a frog an insect? If not, what group does it belong to?

What are marsupials?

Are slugs just snails without a shell?

What are monotremes?

Are dolphins and whales mammals? Why?

Are insects warm-blooded? Do they have hearts?

Choosing a site

Although in many cases the site will have been decided, the more ownership children have in planning and conducting the visit, the more engaged they will be. Discuss the site that has been chosen or discuss with the children what site would be most useful to help them find answers to their inquiries. For instance, if the children are exploring the tools used by people living at a particular time, they may decide to visit a museum or a retirement village to interview people. If exploring the reasons that animals are becoming endangered, they might visit a zoo or talk to wildlife rescue staff. If researching native plants, light and shadows, or water conservation, the school itself may be an ideal site.

Teacher preview

Become familiar with the site. Take photographs to show your class to stimulate their curiosity. Encourage them to visit the site's website. Places such as museums or zoos and even parks change frequently. So, don't rely on a visit from years ago. Worksheets in your filing cabinet can be outdated. Keep an eye out for opportunistic, local learning adventures and environmental impacts. Excavation near the school might have exposed water pipes: 'Where do they go?'; 'What are they made of and why?' Diggings may have revealed the structure of a road: 'What are the different layers?'; 'Where do the materials come from?'

Identify concepts and skills your children need to gain the greatest benefit from the visit. These need to be developed or refreshed at school before you leave. If visiting a venue that has its

Snapshot 12.1: Year 6 investigate meerkats at the zoo

The term theme for this class of Year 6 students was Africa. In science they focused on African animals. The teacher organised two visits to the local zoo that has an African savannah exhibit. The first visit was planned for initial orientation and familiarisation with the African savannah animals and the second was planned for the students to collect data. Groups of students selected different animals in which they were interested. One group of students chose the meerkat because they had recently seen a television series called *Meerkat Manor* (Hawkins, 2005) on the television. Before the students went on the excursion to the zoo, they were involved in writing a scientific report about their selected animal. They used information from resources such as books, video and the internet to write their scientific report. Figure 12.3 shows the report plan on meerkats prepared by one group of students.

　　Based on their report and their initial visit to the zoo, the students were encouraged by the teacher to develop a plan to collect observation data that would help them to understand their animal better. The group of students who were interested in meerkats wanted to know more about what meerkats do all day. With the help of the teacher they decided to watch one meerkat for 15 minutes and record its behaviours. Based on their school-based research, they already knew that meerkats spend some part of the day on sentry duty and they also forage for food. From the first visit to the zoo, they also discovered that meerkats spend time digging and resting on their tummies. The students used these behaviours to design a data chart to record their observations when they returned to the zoo. You can see the students' completed data chart in Figure 12.4.

　　The students completed their observation of a meerkat on their second visit to the zoo. The students had difficulty deciding which behaviour the meerkat was involved in during each minute of the observation because it kept changing behaviours. They decided to tick the behaviour they thought the meerkat did the most during each minute. One student was responsible for the timing, another was responsible for watching the meerkat and the third student was responsible for recording the data on the data chart. When back at school, the students constructed a graph of the meerkat's behaviour (Figure 12.5) and included it in their final report.

Figure 12.3: An example of a scientific report plan about meerkats prepared by Year 6 students prior to a visit to a local zoo

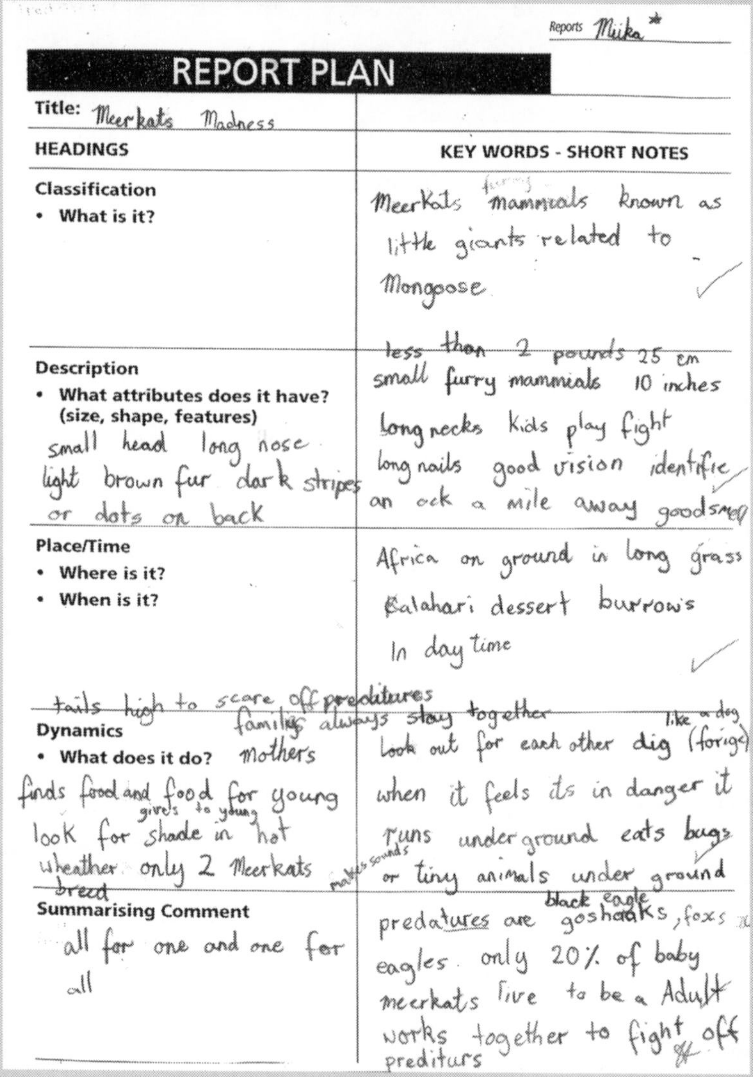

Reports *Mika* *

REPORT PLAN

Title: *Meerkats Madness*

HEADINGS	KEY WORDS - SHORT NOTES
Classification • What is it?	Meerkats *furry* mammals known as little giants related to Mongoose.
Description • What attributes does it have? (size, shape, features) *small head long nose* *light brown fur dark stripes* *or dots on back*	~~less than 2 pounds~~ 25 cm small furry mammals 10 inches long necks kids play fight long nails good vision *identifie* an ock a mile away *good smell*
Place/Time • Where is it? • When is it?	Africa on ground in long grass Kalahari dessert burrows In day time
~~tails high to scare off~~ *predatures* *families always stay together* **Dynamics** • What does it do? *mothers* *finds food and food for young* *gives to young* *look for shade in hot* *wheather only 2 Meerkats* *makes sounds* ~~breed~~	*Like a dog* Look out for each other dig (forige) when it feels its in danger it runs under ground eats bugs or tiny animals under ground
Summarising Comment *all for one and one for all*	predatures are *black eagle* goshawks, foxs eagles. only 20% of baby meerkats live to be a Adult works together to fight ~~off~~ prediturs

Figure 12.4: An example of a data chart prepared by Year 6 students to record their observations of a meerkat at a local zoo

Zoo Observation Data Chart – Meerkat's Behavior   10:00 – 10:15am

~minutes~

Behaviors ↓	1	2	3	4	5	6	7	8	9	10	11	12	13	14	15	Total
Resting	✓	✓			✓									✓		4
Sentry Duty			✓	✓	✓							✓	✓		✓	6
Foraging						✓	✓		✓	✓						4
Digging								✓								1

✓ – The behavior shown the most in the minute

Figure 12.5: A graph of observation data collected by Year 6 students at a zoo visit. The graph was included in the students' final report.

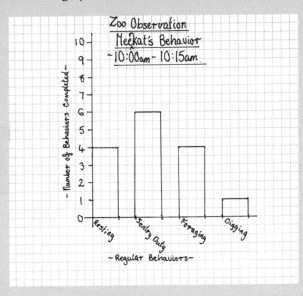

own staff, discuss your particular needs and the topics you will be exploring with them. Read any material supplied by the site and ensure all teachers accompanying the group have a copy.

Plan to be considerate to other people using the site. Plan for safety. State Departments of Education have policies on excursions. Some require pre-visits, risk assessments and/or police checks. Visit your Department of Education's website for further details and advice.

Helping students to think about how to learn on the visit

Discuss the place you will be visiting with your students. The novelty of being at a site away from the classroom may be motivating, but it also can be distracting. Talk to the children about the purpose of the site. Who works there or who looks after it? Look at a map of the site. What people might they meet there? Think of novel aspects that may distract the children. For example, when children visit a natural history museum they are often more intrigued about how animals are stuffed than about their topic. A visit to a heritage site may raise many questions about authenticity. If you are unsure about potential distractions, talk to staff at the venue before you visit.

Talk to your students about how to use the site to find out things they are interested in. Snapshot 12.1 provides an excellent example of students planning to collect information by carefully observing an animal at the zoo. They may need to think about different ways of learning in different parts of a site. In one area it may be appropriate to crawl under a turtle shell to get a feel for the turtle's perspective, the weight and touch of the shell. In others, objects may be less robust and learning may depend on reading, looking and discussing. Advise the children to ask questions at demonstrations. They will need little encouragement to use hands-on exhibits but suggest they read instructions first to get the most from them. Encourage them to draw what they see and record their ideas, questions and answers. They should determine how to do this before the visit.

Planning with the students

You and your students should have a clear overall learning purpose. Together, write a one-sentence statement of why the class is going on this trip. Write this sentence in terms of a learning outcome. Having a clear learning goal for the day not only focuses teachers and students, it also lets parents know why the trip is useful. Discuss and plan the following organisational details with your class.

- When do we leave and return?
- What do we take?
- When and what will happen for lunch?
- Who will we meet at the site?
- Who is going with us?

Discuss with the students where to go within the site. Plan observations the children will make and information they will seek. Plan how to record observations, findings and information. Take into account the report they plan to present: a poster, a web page or an oral presentation, for example.

An information sheet for students and parents

Once agreed, provide children and accompanying parents with an information sheet including such things as the learning purpose, the name of the site and a schedule. It is a good idea to include a plan or map of the site with the information sheet, to identify toilets, eating places and other important locations such as a meeting place in case someone gets lost. The information sheet should provide the students and their parents with clear and concise information on the purpose and conduct of the day. Figure 12.6 provides a pro forma you can use to develop an information sheet for an out-of-school learning experience. You can attach a tear-off permission slip to the information sheet, but the information sheet itself should be taken on the trip. This eliminates all the 'When are we having lunch, Miss?' questions. It also gives some autonomy and ownership to the groups of students and the accompanying parents over the conduct of their visit.

Learning groups and group leaders

Before the visit, confirm the learning groups that were developed in class from the children's specific subtopic interests. (Ideally, there shouldn't be more than eight children in a group.) Allocate a parent helper or teacher to each group as group leader and tell the children

Figure 12.6: A pro forma information sheet that can be adapted for your out-of-classroom site visit

---

OUR VISIT TO _____ (site name) _____
We are going on _____ (date)_____
We are going to learn about _____
_____
_____

We leave school at _____ (time) _____
We arrive at _____
We have lunch at _____
We leave the site at _____
We will wear (school uniform and take our lunch in a soft bag).
We will work in groups. My group leader's name is _____
and the other members of my group are _____
_____
_____
_____
_____

I especially want to find out about _____
_____
_____

Site map

---

who this is. If possible, hold a meeting with the group leaders before the visit. Alternatively, talk to them on the phone before the visit, the morning of the visit, or on the way in the bus. Explain the visit's learning purpose to the group leaders and their role in facilitating students' learning. Explain that this is not to tell the students the answers to their questions but, rather, to act as a learning partner, guiding the students to find their own answers in their own ways. Suggest types of questions group leaders could ask to guide the children's learning and remind them that exhibits are often designed for children to learn through play (Watson, Aubusson, Steele & Griffin, 2002). In addition to the general information sheet, group leaders could be given a second information sheet outlining their role during the site visit.

DURING THE VISIT

The visit should be guided by the students' questions and the plans the class has made to address these. For example, in a museum, a question about what penguins eat and how they keep warm will require a visit to a bird exhibit, a question about who discovered Australia might lead to a gallery about indigenous Australians. At an outside site, a question about the animals that live in the local bush might result in a bushwalk, a tape recording of bird calls, photographs snapped of darting lizards and quick sketches of butterflies.

Orientation to the site

When you arrive, remind the children of the purpose of the visit, and help them to become physically oriented to the site—where things are, what the boundaries are. Relate this to their map. If practical, let them have a general look around to get to know the place.

Allowing learning choice

If lines of inquiry have been well-established through explorations at school before the site visit, student activities at the site will typically be well-directed and targeted. The children want to use the site to find things out. Parent helpers stay with their group throughout the visit as each group works independently, moving where and when they wish, having a break when they wish. Within reason, the children decide when and where they will move, not the adult. Each group reviews its learners' questions and chooses where to go to investigate selected questions. Some groups may prefer first to wander a site, viewing places of special interest.

Group leaders (teachers or parents) are encouraged to make suggestions about how children may explore a site or help to focus the children's learning by asking key questions, such as:

- Why do you think this is interesting?
- What is this?
- Why do you think it is like it is?
- What makes you say that?
- How might it be used?
- Who might have used it?
- Where might we find out about that?

Encourage children to gather further questions, at the site, for later investigation. The visit is not intended to be a learning endpoint, but part of an ongoing learning program.

Recording

Before the visit, children should have planned what information they need and how they might record it. Figure 12.4 in Snapshot 12.1 gives an example of how students can prepare their own data chart to take with them and record information on a site visit. The children also should have thought about how they will report to the rest of the class as the proposed final product will influence the types of records made. Invite the children to choose how they will record their findings. They could:

- photograph or video points of interest;
- carry a tape recorder and talk about interesting things;
- write information; or
- draw objects related to their area of inquiry.

Some children will prefer not to record on site, but will talk about or record what they learned back at school.

If the site is local or in school grounds, records made at regular intervals can reveal interesting information. For example, certain birds and insects appear at the same time each year. Why do they come and where do they go? Why do some *Banksia* produce flowers in winter and what eats the nectar? Some questions can be asked in advance, stimulating student investigation and learning through data collection at the site. Other questions will be stimulated at the site and best answered with resources in the classroom.

AFTER THE VISIT: BACK AT SCHOOL

Linking site and school investigations

Have a whole-class discussion about the visit with your students. In addition to asking them what they liked or did, you should ask: What did they find out? What puzzled them? Were things different from what they expected? How and why? What new questions were raised? What questions were they unable to answer?

Display the students' records in the classroom. Link what they were able to find out with work they were doing before their visit and discuss how their new information informs their overall understanding. Figure 12.5 in Snapshot 12.1 shows how students were able to develop the data they collected about meerkats into a graph and report after the site visit.

Completing investigations and reporting

Children should be given time to incorporate the information gathered at the site into their school-based learning project and given opportunities to do further research on questions or ideas stimulated by the visit. Finally, the children should bring their learning together into a report or presentation that can be shared with the rest of the class, and possibly parents and other school members. For example, these reports could be verbal or written, or take the form of posters, videos, books, games, 3-D displays, plays or poems.

SUMMARY

The key to good learning beyond the classroom is to integrate it with a classroom-based learning unit. Embedding the visit in a school topic makes the purpose of the visit clear and this purpose is most appropriate when developed in collaboration with the students. It should be emphasised that the site is a place for gathering information—a means of learning, not an end in itself. Learning outside the classroom may involve strategies that are different from classroom learning and the range of possible methods for finding information should be explored and discussed before the visit. It is best if small groups of children have choice in selecting a particular aspect of the topic to explore. This gives the children considerable ownership of their learning, allowing them to determine where, when and what they do within a scaffold provided by the teacher. Small group interaction is encouraged and teachers and parents are active participants, with the students, in the learning process.

REFERENCES

Biddulph, F., & Osborne, R. (Eds.). (1984). *Making sense of our world: An interactive teaching approach*. Hamilton, NZ: Science Education Research Unit, University of Waikato.

Falk, J., & Dierking, L. (2000). *Learning from museums*. Walnut Creek, CA: AltaMira Press.

Griffin, J. (1996). *SMILES: School–museum informal learning experiences*. Sydney: University of Technology, Sydney, NSW.

—— (2004). Research on students and museums: Looking more closely at the students in school groups. *Science Education, 88*(Suppl. 1), S59–S70.

Griffin, J., & Symington, D. (1997). Moving from task-oriented to learning-oriented strategies on school excursions to museums. *Science Education, 81*(6), 763–780.

Hawkins, C. (Creator). (2005). *Meerkat Manor* [Television series]. Sweden: Animal Planet.

Hein, G. (1998). *Learning in the Museum*. London: Routledge.

Lucas, B. (1997). Learning through landscape: The importance of school grounds. *Australian Journal of Environmental Education, 13*(2), 85–88.

Watson, K., Aubusson, P., Steele, F., & Griffin, J. (2002). A culture of learning in an informal museum setting. *Journal for Australian Research in Early Childhood Education, 9*(1), 125–137.

CHAPTER 13
SCIENCE TEACHING AND LEARNING IN THE EARLY CHILDHOOD YEARS

Christine Howitt
Curtin University of Technology, Western Australia
Mary Morris
Edith Cowan University, Western Australia
Marj Colvill
Perth Primary School, Tasmania

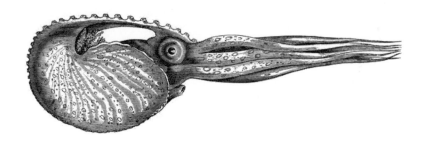

OUTCOMES

By the end of this chapter you will be able to:
- describe the characteristics of effective science teaching and learning in the early childhood years;
- relate these characteristics to early childhood science learning experiences; and
- describe the role of the teacher in effective science teaching and learning in the early childhood years.

INTRODUCTION

Young children are natural scientists, because of their immense curiosity and thirst for knowledge. Science is a part of their everyday world. They question everything around them, observe objects in minute detail and will happily manipulate objects again and again to see if the same thing happens. They are continually trying to make sense of their own personal worlds. This same scientific curiosity can be brought into an early childhood centre or classroom. (Throughout this chapter we are using the international definition of early childhood education: from birth to eight years of age.) By tapping into this curiosity in an appropriate manner, it is possible to empower young children with scientific understanding, basic scientific processes and positive attitudes towards learning science that will stay with them through their future schooling (Smith, 2001).

The role of the early childhood teacher is to nurture this curiosity and thirst for knowledge by providing opportunities, in a safe and caring environment, for young children to explore, question, observe, discover and share their wonder of the natural world. To do this, the early childhood teacher must acknowledge the importance of play as a platform for both learning and practising the basic process skills of science (Arthur, Beecher, Death, Dockett & Farmer, 2005), and provide opportunities for such 'learning through play' experiences.

The purpose of this chapter is to describe the characteristics of effective science teaching and learning in the early childhood years, and the role of the teacher in that process. This will be achieved through the introduction of three case studies, and subsequent detailed description of various characteristics from these case studies. Each case study will highlight a different role of the teacher: facilitator, planner or observer. A simple model of effective science teaching and learning in the early childhood years will then be presented, based on these characteristics.

CASE STUDY 1: WHAT WAS FOUND IN SHREK'S SWAMP?

There were giant green footprints leading into the Year 2 (6–7-year-olds) classroom, and towards the mat area. The sign on the door also gave away that something different was happening in science today. The sign said, 'OGRE—BEWARE!' and there was also a picture of Shrek on the sign, just like in the movie.

The children all sat down on the mat area and their teacher, Mr J, welcomed them to Shrek's swamp. There were squeals of delight from the children, and Mr J had to remind them to quieten down. Mr J started by telling the children that he had found a strange substance in Shrek's swamp, and he required the children's expertise to work out what it was. The children were extremely interested in what their teacher was saying. Mr J explained to the children they would work in their normal groups at the tables, and they had to answer six questions (already written on the whiteboard) in relation to the strange substance that he had found in the swamp:

- What does it feel like?
- What does it look like?
- What does it smell like?
- What does it sound like?
- How does it move?
- What does your group think it is?

The children quickly moved to their groups. Mr J placed a container of the strange substance (cornflour and water with green food colouring) on each of the tables, which were already covered in newspaper. Some of the children immediately placed their hands into the containers, eager to feel this strange substance. Others were hesitant, and waited for Mr J to show them what to do.

Once Mr J had seen that all the children were engaged with the strange substance, he (along with two parent helpers) circulated around the groups. At each table Mr J would ask 'I wonder what would happen if we hit it hard?', and he and the children did. He also asked 'I wonder what would happen if we placed it on our fingers?', and he and the children did. Mr J also asked the children questions relating to those that had been written on the board. There was considerable discussion within each of the groups about how to describe the strange substance. Many imaginative and highly descriptive words were used.

After 20 minutes, Mr J gained the children's attention and the class then brainstormed the answers to the questions on the whiteboard. The result of the children's brainstorm is shown in Figure 13.1.

Figure 13.1: Results of the children's brainstorm on what was found in Shrek's swamp

What does it feel like?

wet, slimy, gooey, stretchy, smooth, strange, runny, icky, dry, crumbly, hard when you hit it, soft when you stir it

What does it look like?

playdough, slime, flubber, goop, a marshmallow that you have sucked, shiny, icing sugar, alien's blood

What does it smell like?

nothing, flour, bread, old

What does your group think it is?

Shrek's ear wax, Fairy Godmother's magic potion, Shrek's toothpaste, Princess Fiona's nail polish, Shrek sneezed

What does it sound like?

gloop, whack, drip, it doesn't make a sound

How does it move?

fast, stretchy, oozes, weird, drippy, spreads out as if it were alive

At the end of the lesson, when the children were sitting back on the mat area, Mr J asked if they had any questions or wanted to know anything else about the strange substance. Most children put their hands up and asked questions. Many children wanted to know what it was made of, and could they make some and take it home. Others wanted to know if it could be another colour, preferably their favourite colour. Some children asked if they could eat it. Mr J wrote these questions down, and assured the children that all these questions would be answered over the coming weeks, along with other investigations into this strange substance.

Analysis of Case Study 1

This single lesson case study highlights many of the characteristics of effective science teaching and learning in the early childhood years. Science in the early childhood years should be simple, inclusive, hands-on, student-centred; it should allow for exploration through the process skills, be integrative, allow for questioning and be play-based. Each of these characteristics is discussed below in relation to this case study.

Simple   Science in the early childhood years should be simple. Young children do not need complex activities or investigations to help them learn about science. The simplest and most basic activities, as illustrated here, can develop children's scientific understanding, scientific processes and positive attitudes towards learning science. Activities such as blowing bubbles, pouring water, digging in the sand or exploring the playground can be used to create meaningful science learning experiences for young children.

Science should also be simple for the teacher. This means using everyday objects available in supermarkets, or using the playground or schoolyard, to help young children explore science. Specialised equipment might promote an image of science for the elite. The use of everyday objects makes science more accessible and easier to prepare for teachers. Taking children on a school walk, carrying a basket to collect any interesting objects they see, and a digital camera to record those objects they cannot collect, is another example of the simplicity of early childhood science. Young children see science everywhere. Early childhood teachers should follow their example.

Inclusive   Science in the early childhood years must be inclusive. This means that all children should have the opportunity to participate and to experience success in their science activities. Case Study 1 readily illustrates this, as all the children were provided with the opportunity to explore the strange substance and describe it. In the class brainstorm, all answers were accepted. There was no right or wrong answer. Such inclusivity increases children's confidence in and enjoyment of participating in science. Inclusivity also refers to the shared nature of the children's learning experiences. In this case study the children were sharing their language, ideas, prior knowledge and enthusiasm with the other children in their group and in the class. The sharing of learning experiences allows the children to both hear and appreciate alternative views and ideas, and discuss them within a safe and supportive environment.

Hands-on   Young children learn in a concrete manner. Hence, their science learning experiences need to be very hands-on and visual. Manipulating the strange substance in Shrek's swamp provided a rich, hands-on experience for the children. Similar hands-on experiences, as described by Smith (2001), include developing a worm farm, making compost, growing vegetables, observing birds and animals in the playground and exploring magnets.

Student-centred   Early childhood science should be student-centred. The strange substance was introduced to the children within the context of Shrek's swamp. That term, the children had been studying fairy tales and had already watched the movie *Shrek* in class. By using the giant green footprints leading into the classroom, along with the sign on the door, the teacher immediately motivated the children and gained their interest. Such a student-centred approach to science means selecting topics that the children are interested in or are curious about.

Exploration through the process skills   Exploration within early childhood science can be developed through process skills (Fleer & Hardy, 2001). The basic science processes of observation, communication and classification can be readily incorporated into many learning experiences. In this case study, the children had to use their senses to describe the strange substance, share their descriptions with the class and then attempt to identify the substance. Within their description of what it looked like, most children were attempting to classify the strange substance based on their prior knowledge and experiences of the movie.

Integrative   Early childhood science readily integrates with the other learning areas. In the case study there was a high level of integration of science with literacy, as can be seen from Figure 13.1. Developing the process skills allowed language to become a prominent part of the science learning experience. Such literacy could be further developed and supported through the use of a word wall (Australian Academy of Science, 2005), or asking each student to draw, describe and name the strange substance (see Chapter 10 for more information about the literacies of science).

Questioning   Questioning is a pivotal component of early childhood science. Questioning allows children to satisfy their own curiosity, and provides the teacher with a window into their thinking processes and imagination. In Shrek's swamp, the children were provided with many opportunities to question themselves, each other and the teacher about the strange substance. The questions at the end of the lesson formed the basis of the remaining lessons for the term. Such an interactive approach to teaching science ensures that the children remain interested and motivated, as they are doing what they want to do and are answering their own questions (Fleer & Hardy, 2001).

Play-based   Learning through play occurred within Shrek's swamp as the children were allowed time to play with the strange substance. While they were playing the children were encouraged to think about the answers to the questions. Playing allowed the children the freedom to manipulate and observe the strange substance to satisfy their own curiosity about what it was and how it worked.

CASE STUDY 2: BILLY'S SNAIL

Billy brought a snail into his pre-primary (4–5-year-olds) classroom. All the children were very excited, and gathered around Billy, wanting to have a closer look at the snail. Billy was so squashed that he was afraid of dropping the screwtop jar which was the very temporary home of the snail.

The teacher, Miss P, told the children that today they would begin investigating Billy's snail. Miss P asked the children what things the class could investigate about Billy's snail. The class came up with the following suggestions: watch him, draw him, feed him, paint him, name him, look after him, and how do we know it's a him?

From these suggestions, Miss P asked the class what Billy's snail would need to live happily in the classroom for a short period of time. The class came up with the suggestions presented in Figure 13.2.

Figure 13.2: Possible starting points for investigations into Billy's snail and its needs

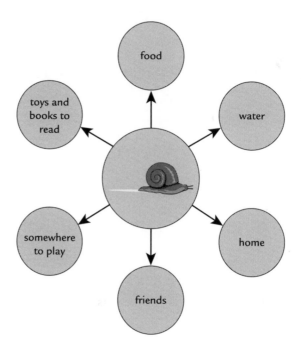

The class decided that the first thing they wanted to do was to build a home for Billy's snail. The snail's jar was placed in a cool spot and the class went outside on a collection trip around the playground. The collected items were brought back to the classroom and spread out on sheets of newspaper. The items were sorted into groups (rocks, stone, moss, leaves and sticks) and their suitability for the snail's home was discussed. The children realised that they didn't know enough about snails and decided they had to find out more information using the internet (with Miss P's help). From this new information, and their prior experiences, the children used an old aquarium with a secure lid and the materials they had collected to make Billy's snail a new home.

In subsequent lessons the class investigated the different foods that snails like to eat, how quickly snails grow, competed in snail races, sketched and painted snails, and found out about snails in the food chain and about other types of snails.

Analysis of Case Study 2

This case study highlights many of the characteristics that have already been discussed in the first case study: it is simple, inclusive, hands-on; it allows for exploration through the process skills, is integrative, questioning and play-based. However, it also highlights two characteristics that were not specifically covered in the first case study: student-centred planning and allowing time and depth for exploration.

Student-centred   A student-centred approach to science teaching and learning draws on young children's natural curiosity and places them at the centre of their learning. Miss P immediately recognised the children's curiosity about Billy's snail, and used that curiosity to develop an integrated program for the term. By listening to the children and finding out what they were interested in, Miss P was showing the children that she valued their ideas, interests and questions. These were subsequently used, via Figure 13.2, to develop a teaching program for the term. A student-centred approach is both motivating and relevant to the children as they are answering their own questions and researching what interests them.

Allow time and depth for exploration   Science topics in the early childhood years should provide sufficient time and depth for the children to form concepts from multiple experiences (Smith, 2001). Case Study 2 clearly illustrates this, allowing a whole term of work with in-depth learning experiences to allow the children to gain a more thorough understanding of both snails and the investigation process. Children should be given time to explore; freedom to experiment with materials, activities and ideas; and many opportunities to interact with other children and adults. Allowing both time and depth for exploration encourages the transference of conceptual understanding based around simple science investigations.

CASE STUDY 3: LUKE'S TRUCK

At 9 months of age Luke was given a new tip truck and a 4-wheel drive utility truck. Mrs L observed her grandson playing with the trucks on the wooden floor in the lounge. Luke pushed the 4-wheel drive truck backwards, tipped it over and picked it up by the front wheels. He placed one wheel in his mouth. Luke then put the truck down upright on the floor and pushed it backwards along the floor.

Luke then turned his attention to the tip truck. He grabbed the tray of the truck, making it tip. He then pushed the truck backwards. Luke pulled the truck towards himself, and then pushed it away from himself. He tipped up the tray and pushed the truck forwards along the floor. Luke then pulled the truck towards himself, bent down and bit the truck's plastic driver. He tipped up the tray again, held it back with his right hand, and put his head down into the bottom of the truck. Luke put the tray down and started placing plastic construction blocks into the tray. He then tipped the blocks out. Luke then pushed the truck along the floor.

Five months later Luke was playing outside in the courtyard with the same two trucks. He picked up the utility truck by the roof and then put it down. He laughed and pushed the utility backwards away from himself. Luke then picked up the tip truck by the tray, and pushed it away from himself. He returned to the utility, pushing it backwards. Luke placed some pebbles in the back of the utility. He picked up the tip truck by the roof and placed it on the ledge above, with the front towards himself. Luke turned the utility around so that it was facing forwards, and made a noise like a truck moving: 'brrrm'. As he pushed the utility forward, he continued making the noise.

Analysis of Case Study 3

This case study highlights the early signs of developing science in very young children. Two characteristics will be discussed in detail: exploration through the senses and the use of simple equipment.

Use of the senses In the 0–3-year-old developmental stage, very young children are continuously using their senses to explore the world. In particular, they use their sense of taste and their tactile sense to explore the materials from which objects are made and their visual sense to recognise familiar objects. Luke's initial exploration of his trucks highlights this use of the senses. Over time, as Luke became more familiar with his trucks, he started to explore them using other materials, such as the pebbles. He also developed an association between his trucks and those experienced outside his home. By imitating the sound of the trucks, he began incorporating the auditory sense into his play.

Simple equipment Young children can be exposed to a wide range of science experiences and concepts using simple equipment. Playing and interacting with toys made from different materials provides many opportunities to discover the properties of those materials. Today's plastic toys are made of materials to emphasise their appearance, such as bright or fluorescent colours, and a glossy finish. Some baby toys have mechanisms that relate to the use of energy—balance toys such as mobiles, pop-up toys, pull-cord toys with springs, or press button toys that move a lever. As children get a little older they experience toys with simple machines, such as the pulley, or cars with wheels and axles. In this case study Luke was exploring the concept of wheels, and observing the movement of the trucks' wheels on the wooden surface of the floor and concrete slabs outside.

Everyday simple equipment found in the home can provide ideal science learning experiences for young children. The plastic container cupboard in the kitchen can provide hours of fun for a toddler as they bang a wooden spoon on plastic containers and make the sound of a drum, or roll a plastic cup down a plastic tray that is being used as a ramp. The manipulation of simple, common, everyday items provides young children with an experience of many science concepts.

ROLES OF THE EARLY CHILDHOOD TEACHER

These case studies highlight the different roles of the teacher when teaching science in the early childhood years: facilitator, planner and observer (Fletcher, Edelman & Sampson, 2000). Each of these is discussed separately below.

Facilitator

In Case Study 1, Mr J was a facilitator of learning. He provided the time, opportunity and the resources required for the children to have rich and stimulating science learning experiences. He also established a meaningful context for the learning experiences through Shrek's swamp. Further, Mr J was active in supporting, guiding, responding to and extending the learning of the children (Fletcher et al., 2000). Mr J also took on a variety of other roles including that of modeller, questioner and co-investigator.

Modeller Modelling is an important part of early childhood teaching and learning, and can be used in a variety of ways. Modelling can be used to demonstrate basic skills that are required for the learning experience. Young children are still developing many basic skills, and may require assistance to complete an activity. For example, Mr J made a point of exploring the strange substance with each group of children, showing them how to hit it and how it could flow off their hands. At the same time, Mr J was also modelling his enthusiasm for science. This is an increasingly important role, as teachers' attitudes towards science can have a marked influence on their children's attitudes (Koballa & Crawley, 1985). Mr J made sure the entire learning experience was positive, engaging and fun for the children, and that he reflected those very attributes. His enthusiasm became the children's enthusiasm.

Questioner Being an effective questioner is important in science. Questions can be used to establish prior knowledge, direct the learning experience, help children to express their thoughts and ideas, and plan for future lessons. By asking 'I wonder what would happen if . . . ?', Mr J was directing the children to particular experiences he wanted them to observe. He was also using the questions to extend the children and their thinking. However, the open-ended nature of the questions still let the children make the discoveries themselves.

Co-investigator Early childhood teachers of science are also co-investigators. They explore, discover and are actively involved with the children during investigations. Co-investigators must be both modellers and questioners for the children. As a co-investigator, the teacher must work at the same level as the children, taking their cues from the children, asking questions, giving instructions, or responding to comments. Working in this role allows children to see that teachers do not have to know the answer to every science question, but are prepared to find out with the children. It also allows the children to see that everyone can be a scientist.

Planner

In Case Study 2, Miss P was a planner. She allowed the children's interest in Billy's snail to guide her term planning. Miss P also allowed for both planned and unplanned learning experiences, so that direct and indirect learning could occur. For example, she planned on taking the children out of the classroom to collect items for the snail's home. However, she did not know what the children would collect. Miss P's approach to planning was also integrative and flexible. Her flexibility allowed time for pretending, imagining, creating and improvising within the children's play and learning. Within her planning, Miss P allowed the children to become responsible for their own learning. Her student-centred approach to planning empowered the children.

Observer

In Case Study 3, Mrs L was an observer. She observed Luke at play to understand his interests, physical skills level and knowledge in relation to the object he was exploring (Fletcher et al., 2000). Observing assisted Mrs L to identify Luke's strengths and learning patterns, as well as his most immediate needs (Fletcher et al., 2000). These types of observations can provide a major input into the planning phase of a teaching program as the development of the child is being closely monitored, and the teacher can plan what materials and equipment to place into the child's immediate environment. The case study shows that Luke has made the association between the purpose of the truck, the noise it makes and how it moves. This information can then be transferred to a different context, perhaps with another transport toy. In her role as observer, Mrs L was allowing Luke to use a discovery approach to his learning without intervention, and allowing him time to play and explore objects within his environment.

MODEL OF EFFECTIVE SCIENCE TEACHING AND LEARNING IN EARLY CHILDHOOD

These three case studies together allow a simple model of effective science teaching and learning in early childhood to be developed. The first case study provided a detailed single science lesson, the second illustrated a planning approach to science and the third illustrated science in the very early years. All three case studies had many similar characteristics.

A model of effective science teaching and learning in early childhood is illustrated in Figure 13.3. This model uses the analogy of an icecream cone. The cone represents the early childhood teacher, while the scoops of icecream represent the different characteristics developed from the three case studies.

In this analogy, the cone supports the scoops of icecream, just as the teacher supports the science teaching and learning process. The cone has three sides, representing the three major roles of the teacher—facilitator, planner and observer.

Each scoop of icecream represents a different characteristic of effective science teaching and learning in the early childhood years.

Figure 13.3: Icecream cone model of effective science teaching and learning in early childhood

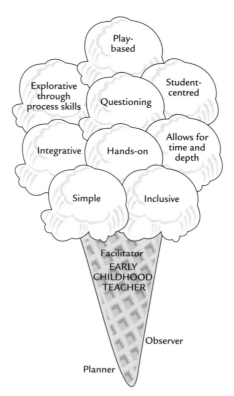

Just as eating one scoop of icecream is not very satisfying, teaching science using only one characteristic provides a limited experience. However, teaching science with all the scoops together provides for motivating and meaningful learning experiences. Similarly, when young children are given the opportunity to eat all the scoops of icecream, they become involved in exciting and rewarding science.

SUMMARY

Teaching science in the early childhood years should take advantage of children's natural curiosity and their thirst for knowledge. Such science should be simple, inclusive, hands-on, student-centred, allow for exploration through the process skills, be integrative, questioning and play-based and provide sufficient time and depth for exploration. The teacher can take on different roles when teaching science in the early childhood years, the three main ones being facilitator, planner and observer. As a consequence of teaching science with this model, young children should be engaged in quality learning experiences that will empower them with scientific understanding, basic scientific processes and positive attitudes towards learning science.

REFERENCES

Arthur, L., Beecher, B., Death, E., Dockett, S., & Farmer, S. (2005). *Programming and planning in early childhood settings* (3rd ed.). South Melbourne, Vic.: Thomson.

Australian Academy of Science. (2005). *Primary connections. Weather in my world. Early Stage 1.* Canberra: Australian Academy of Science.

Corrie, L. (2000). *Learning through play. 1. Curriculum and learning outcomes.* Perth, WA: University of Western Australia.

Fleer, M., & Hardy, T. (2001). *Science for children* (2nd ed.). Frenchs Forest, NSW: Pearson Education.

Fletcher, J., Edelman, L., & Sampson, L. (2000). *Learning through play. 2. Teachers' role in children's play.* Perth, WA: University of Western Australia.

Koballa, T.R., & Crawley, F. (1985). The influence of attitude on science teaching and learning. *School Science and Mathematics, 85,* 222–232.

Smith, A. (2001). Early childhood—a wonderful time for science learning. *Investigating, 17*(2), 18–20.

INDEX

From the editors of *The Art of Teaching Primary Science*, a resource for secondary science teachers:

The Art of Teaching Science
edited by Grady Venville and Vaille Dawson

Science teaching is an art that requires a unique combination of knowledge and skills to engage students and foster their understanding. An invaluable resource for secondary and middle school science teachers, pre-service teachers and science teacher educators, this book is a thorough introduction with its finger firmly on the pulse of the challenges of the science classroom.

The Art of Teaching Science presents science teaching as a dynamic, collaborative activity and explores recent changes in the theory underlying excellent science teaching. Chapters on planning, teaching approaches, resources and assessment cover the key elements of practical science education. Controversial issues in science, equity and integration across the curriculum are also addressed. Teachers are encouraged to maximise the potential of electronic and traditional resources both in and out of the lab to engage their students.

'. . . a welcome addition to the resources science teachers (including those preparing to teach) have to draw on to improve the quality of the science learning of their students. The authors have done an excellent job of interpreting how a number of the exciting ideas and teaching strategies that we know from recent research can lead to better learning of science.'
Peter J. Fensham, Emeritus Professor of Science Education, Monash University

'The editors have assembled a group of enlightened science educators to share their passion for teaching. Each chapter highlights ways in which teachers can make their teaching effective, exciting and satisfying.'
Gary Thomas, President, Australian Science Teachers Association

ISBN 978 1 74114 217 4